Copernicus Books

Sparking Curiosity and Explaining the World

Drawing inspiration from their Renaissance namesake, Copernicus books revolve around scientific curiosity and discovery. Authored by experts from around the world, our books strive to break down barriers and make scientific knowledge more accessible to the public, tackling modern concepts and technologies in a nontechnical and engaging way. Copernicus books are always written with the lay reader in mind, offering introductory forays into different fields to show how the world of science is transforming our daily lives. From astronomy to medicine, business to biology, you will find herein an enriching collection of literature that answers your questions and inspires you to ask even more.

José Cordeiro • David Wood

The Death of Death

The Scientific Possibility of Physical Immortality and its Moral Defense

José Cordeiro
Madrid Singularity
Madrid, Spain

David Wood
London Futurists
Surbiton, Surrey, UK

ISSN 2731-8982 ISSN 2731-8990 (electronic)
Copernicus Books
ISBN 978-3-031-28926-2 ISBN 978-3-031-28927-9 (eBook)
https://doi.org/10.1007/978-3-031-28927-9

This Springer imprint is published by the registered company Springer Nature Switzerland AG
The registered company address is: Gewerbestrasse 11, 6330 Cham, Switzerland

万事开头难。(All things are difficult at the start.)
Chinese proverb

Science does not know its debt to imagination.
Ralph Waldo Emerson, 1876

This book is dedicated to the first generation of immortal humans. Until now, humanity had been doomed to death. Hopefully, now, thanks to the great technological advances to come in the next few decades, we stand on the threshold between the last generation of mortal humans and the first generation of immortal humans.

Thanks to our ancestors we have arrived here, and our descendants will soon be able to enjoy the extension of lifespans as it was never before possible. We will live in a much better world with an indefinite lifespan, indefinitely young; not condemned to involuntary death, nor indefinitely elderly. Furthermore, we will move from the extension of lifespan to the expansion of life itself by increasing our capabilities and possibilities on this small planet and then far beyond the Earth.

This book is dedicated to the young and the old, to women and men, to believers and unbelievers, to the rich and the poor, to all those around the world who work so that we can finally reach the oldest dream of humanity: The Death of Death. The control of human aging and rejuvenation will soon be a reality. It is our ethical duty to move as quickly as possible toward this noble goal.

The right to life is the most important of all human rights. Without life there is no other right that counts. Now more than 100,000 people die every day, day after day, from diseases related to aging. It is the greatest crime against humanity, against all humans, regardless of race, sex, nationality, culture, religion, geography, or history. We must stop this immense human tragedy. We can prevent it now, and we must prevent it now. It is our moral responsibility, our ethical duty, our historical commitment. We must preserve life to avoid further suffering, to eliminate aging, to dispense with death.

Today the question is no longer whether it will be possible to cure aging, but when it will be possible. The sooner, the better. We are in a race against time, and our common mortal enemy is aging and death. The Death of Death is dedicated to the first generation of immortals that will conquer death.

Preface to the English Edition

I part with this book with deep seriousness, in the sure hope that sooner or later it will reach those to whom alone it can be addressed; and for the rest, patiently resigned that the same fate should, in full measure, befall it, that in all ages has, to some extent, befallen all knowledge, and especially the weightiest knowledge of the truth, to which only a brief triumph is allotted between the two long periods in which it is condemned as paradoxical or disparaged as trivial. The former fate is also wont to befall its author. But life is short, and truth works far and lives long: let us speak the truth.

Arthur Schopenhauer, 1819

We are currently living in a historical crisis, which implies present danger but also future opportunity. Regardless of where the Covid-19 crisis originated, it became a global problem and it required a global solution. This unexpected crisis posed a fantastic opportunity to move forward together as one global family in our small planet, if we avoid the danger of standing divided against our common enemy. Covid-19 represents one of the worst pandemics in about a century, since the 1918–1920 "Spanish flu," when it is estimated that maybe 50 (some historians claim even up to 100) million people died of that virus a century ago.[1] That was an incredible human tragedy, 100 million people died out of a total world population of about 2 billion inhabitants at the time; in fact, more people died of the "Spanish flu" than the casualties in World War I, or even in World War II.

Now, fortunately, thanks to the exponential advances of science and technology, we are much better prepared to solve such a global crisis. In this great planetary war against the little new coronavirus, including its possible mutations, many public and private responses appeared. Governments, large companies, small startups, universities, and even individuals worked arduously to figure out how to control, cure, and eliminate Covid-19. Like in most crises, the costs are immense at the start, but then they are quickly amortized by mass producing the best treatments all over the planet. As another historical reference, let's remember that when HIV and AIDS appeared, it took more than 2 years to sequence the virus, and there were no treatments for several more years. AIDS was even considered the perfect disease because it destroyed the immune system itself and thus basically constituted a death sentence, with no possible escape. After years of international research, the first treatments to control HIV were developed, but the first treatments cost millions of dollars each. Today, however, HIV is treated as a chronic disease with antivirals costing hundreds of dollars in richer countries, and only tens of dollars in poorer countries like India. Fortunately, it is also highly likely that in this decade a vaccine will finally be developed to cure and definitively eliminate HIV.

Although it was difficult to believe in these times of crisis, thanks to the great exponential advances in science and technology, the first antivirals and the first vaccines were developed in just a few months, when normally vaccines took between 5 and 10 years to develop. Thus, the terrible pandemic of this coronavirus will be remembered in the future for having been overcome with unprecedented speed. Covid-19 has become a great lesson for humankind. It has shown that we must work together because global problems require global solutions. There will be new pandemics in the future, but we will be more prepared to overcome them quickly thanks to exponential technologies. It is possible that we will sequence the next pandemic virus in just 2 days, and the next one after maybe in two hours, and not in 2 weeks like it was for Covid-19 now, 2 months for SARS two decades ago, or more than 2 years for AIDS four decades ago.

Not only are the times for fighting these diseases being reduced exponentially, but also the costs are falling drastically. The world will be better prepared for new pandemics and new global challenges such as climate change, wars, terrorism, earthquakes and tsunamis, meteorites, and other space threats, among many future challenges. And hopefully this will be the last great pandemic that all of humanity will suffer!

Covid-19 has been a recent threat to humanity, but the biggest enemy of all is aging and death. Longevity has long been considered one of the major blessings of life, and now we have the possibility, for the first time in history, to defeat aging and death. In fact, the discussion has started in several countries whether aging can finally be considered a disease – specifically, as a curable disease. Longevity advocates have started to be active in Australia, Belgium,

Brazil, Germany, Israel, Russia, Singapore, Spain, the United Kingdom, and the United State, among several countries. Which will be the first country to officially declare aging as a disease? From 2018, the World Health Organization (WHO) has started recognizing age-related diseases, with codes XT9T for diseases with measurable age-related processes and MG2A for diseases made worse by biological aging.[2] Which country will take the lead now and start working on aging as a curable disease?

Curing aging is not just a moral and ethical imperative, it will also be the biggest business opportunity in the coming years. The antiaging and rejuvenating industries are just beginning, while the medical costs of our aging societies keep increasing. Some studies indicate that health costs will double by the year 2050, mostly due to the increasing numbers of older people.[3] This will be particularly dramatic in many advanced countries, where the population is not only aging fast but it will also start declining due to fewer children being born. Economically, this will be very problematic since there will be less and less young people to support more and more old people.

Worldwide, there are already more people over age 65 than under 5 years of age,[4] and this trend will continue. Furthermore, the population will start declining in many countries, like it is already happening in Japan and Russia, for example. According to a recent study published in the prestigious British health journal *The Lancet*, the population of China might drop by half to around 732 million people by the year 2100, and other countries like Germany, Italy, Japan, Russia, and Spain are also projected to drop dramatically.[5] Moreover, while many wealthy countries got rich before getting old, other countries are getting old before getting rich. The results will be dramatic unless these countries can do something about that decline and aging of the population today.

It is estimated that Covid-19 could cost humanity over US$ 11 trillion in terms of GDP during the current crisis, but that future similar pandemics can be avoided this decade by yearly investments of US$ 26 billion, according to a study in the prestigious American journal *Science*.[6] Medical costs around the world are already around 10% of global GDP, and raising fast, partly due to aging societies. Now, let's think about how much aging costs to society, and how much could we save by preventing aging. These are some of the ideas behind the "Longevity Dividend" initiative and explained also in this book.[7] Therefore, we will not only save trillions of dollars curing aging, but we will also avoid incredible suffering and pain to the aging people themselves, their families, and their societies.

Covid-19 might help us to realize that there is nothing more important than health, and that the first human right is the right to life. The world has united against this global pandemic which has inspired the largest vaccination campaign in history to contain the Covid-19 spread after the production of more than nine billion vaccines in just one year. New mRNA vaccines are now being developed against many cancers, malaria, and HIV, which are three completely unrelated medical conditions. What seemed impossible before has become possible now. Hopefully, this crisis will also serve as the opportunity to finally start devoting more resources to curing the mother of all diseases: aging itself. Can we finally fulfill the long-term dream of humanity of "immortality"?

This is the best opportunity for action, the time is now, the place is here, who will lead the way?

José Cordeiro
David Wood

Notes

1. https://www.visualcapitalist.com/history-of-pandemics-deadliest/
2. https://www.thelancet.com/journals/lanhl/article/PIIS2666-7568(21)00303-2/fulltext
3. https://www.fightaging.org/archives/2020/08/the-reasons-to-study-aging/
4. https://www.nationalgeographic.co.uk/history-and-civilisation201907there-are-now-more-people-over-age-65-under-five-what-means
5. https://www.thelancet.com/infographics/population-forecast
6. https://www.statista.com/chart/22378/estimated-cost-of-containing-future-pandemic/
7. http://www.sjayolshansky.com/sjo/Longevity_Dividend_Initative.html

Prologue

Aging, like the weather, does not respect national or ethnic boundaries; it affects each group and subgroup of humanity more or less equally. There is much talk about the disparities that exist in this regard; for example, that although the United States is the country that spends the most on health per capita, it is not even in the group of the 30 countries with the highest life expectancy. However, these statistics should not deceive us, since the disparities are numerically small: life expectancy in the United States is only 5 years lower than that of Japan. It is essential that there be no borders in the crusade against aging. The entire world must unite and devote its greatest efforts to solve this problem, since it is the main challenge facing humanity.

Old age kills many more people than anything else. Aging is responsible for more than 70% of deaths, and most of those deaths are preceded by unspeakable suffering, both of the elderly and their loved ones. Unfortunately, the "war on aging" is not yet rising to this challenge. It is gaining considerable momentum in the English-speaking world, where the greatest efforts are concentrated in Silicon Valley, while centers are also emerging in the rest of the United States, in the United Kingdom, Canada, and Australia. Germany is also coming to the fore, as are Russia, Singapore, South Korea, and Israel. However, other parts of the world are much slower in embracing this field. Asia is especially worrying, as its more populated countries seem to have serious difficulties in understanding that aging is a medical problem, and even more so that it is a solvable problem.

The Death of Death is a visionary book that confronts us with the terrible reality of aging, and its authors are connoisseurs of the subject. In recent years, José Cordeiro has helped to raise the profile of the war on aging in many parts of the world, but his main focus has been, quite rightly, the Spanish-speaking and Portuguese-speaking countries. José has been very successful not only because he himself is both Spanish and Latin American (born in Venezuela from Spanish parents), but also because the level of interest in defeating aging in Spain and Latin America, as far as I can see, is increasing.

The book's co-author is the British technologist David Wood, another well-known anti-aging crusader, who brings a different but complementary perspective. David has transformed the British techno-visionary world with his work as head of several organizations in London. It's hard to imagine a more powerful partnership to give the necessary authority to a book about aging and its (hopefully imminent!) defeat.

Given their extensive international experience, there are no better authors than José and David to advance the cause of longevity globally. They have been immersed in the anti-aging mission for many years, and so they are exceptionally informed, not only about the science of anti-aging research and its latest advances but also about the irrational concerns and criticisms that so often oppose this mission. José and David know the best answers to refute the critics and to convince more people of the benefits of radical life extension.

The first edition of this book was in Spanish (*La Muerte de la Muerte*, Editorial Planeta, 2018) and it quickly became a bestseller, first in Spain and then in several Latin American countries. The second edition was in Portuguese (*A Morte da Morte*, LVM Editora, 2019) and it also became a bestselling book first in Brazil and later in Portugal. The third international edition was in French (*La mort de la mort*, Éditions Luc Pire, 2020), the fourth in Russian (*Смерть должна умереть*, Альпина, 2021), the fifth in Turkish (*Ölümsüz insan*, Nemesis Kitap, 2022), the sixth in German (*Der Sieg über den Tod*, Münchner Verlagsgruppe, 2022), and the seventh in Chinese (永生, People's Publishing House, 2023). Now, with an actualization and updated version, *The Death of Death* will be launching in additional languages, first in Korean, second in Japanese, and later followed by expected translations into Arabic, Bulgarian, Czech, Greek, Hindi, Italian, Persian, Polish, Serbian, Slovenian, Tagalog, and Urdu. Based on its previous success, this book will surely continue revolutionizing the world.

I am convinced that this book will play an important role in the war against aging in the coming decade. I also believe that the authoritative and exhaustive description of this crusade that José and David provide in this excellent book will accelerate this process. Onward!

Aubrey de Grey, PhD
Founder of LEV (Longevity Escape Velocity) Foundation and co-author of *Ending Aging*

Introduction

Death must be an evil and the gods agree; for why else would they
live forever?
Sappho, c. 600 BC

The journey of a thousand miles begins with a single step.
Lao-Tse, c. 550 BC

To be, or not to be: that is the question.
William Shakespeare, 1600

The Greatest Dream of Mankind

Immortality has been humanity's greatest dream since prehistory. Human
beings, unlike most other living beings, are conscious of life and therefore
conscious of death. Our ancestors have created all kinds of rituals related to
life and death since the appearance of *Homo sapiens sapiens* in Africa. Our
ancestors practiced these rituals and created many others during the thousands of years of migration all over the planet. The great civilizations of the
ancient world created sophisticated rituals to celebrate when someone died,
rituals that in many cases were the most important element in the lives of
those who survived them. Let us think, for example, of the rigorous, life-long
mourning processes in many of our societies.

The Quest for Immortality

British philosopher Stephen Cave from the University of Cambridge has written in his bestseller *Immortality: The Quest to Live Forever and How It Drives Civilization*:[1]

> All living things seek to perpetuate themselves into the future, but humans seek to perpetuate themselves forever. This seeking – this will to immortality – is the foundation of human achievement; it is the wellspring of religion, the muse of philosophy, the architect of our cities and the impulse behind the arts. It is embedded in our very nature, and its result is what we know as civilization.

Egyptian funeral rites were very sophisticated. The most important rituals included large pyramids and sarcophagi dedicated exclusively to the pharaohs. The oldest *Texts of the Pyramids* are a repertoire of spells, incantations, and supplications engraved on the passages, antechambers, and sepulchral chambers in the pyramids of the Ancient Empire for the purpose of helping the pharaoh in the underworld and ensuring his resurrection and eternal life. They are a compilation of texts, of ancient religious and cosmological beliefs, written with hieroglyphics on the walls of the tombs, which were used during funeral ceremonies from the year 2400 B.C.

Centuries later, the Egyptians compiled the *Book of the Dead*, which is the modern name of an ancient Egyptian funerary text that was used from the beginning of the New Empire around 1550 B.C. to 50 B.C. The text was not exclusive to the Pharaohs and consisted of a series of magical spells intended to help the deceased overcome the judgment of Osiris, the Egyptian god of death and regeneration, to assist them on their journey through the underworld to the afterlife. Although today they are spoken of as mythologies, religion and Egyptian practices to guarantee immortality were practiced for almost 3000 years, that is, for many more centuries than Christianity or Islam to this day.[2]

In Mesopotamia there are even older documents, created around 2500 B.C. in clay tablets with cuneiform writing. The *Epic of Gilgamesh* or the *Poem of Gilgamesh* is a Sumerian narrative in verse about the adventures of King Gilgamesh of Uruk, the oldest known epic work in the history of mankind. The philosophical axis of the poem is found in the mourning of the King Gilgamesh for the death of someone who was at first his enemy and then his great friend, Enkidu. The epic is considered the first literary work to emphasize human mortality as opposed to the immortality of the gods. The poem includes a version of the Mesopotamian myth of the flood that would later appear in many other cultures and religions.[3]

In China, it seems that emperors were also obsessed with immortality. After conquering the last independent Chinese state in 221 B.C., Qin Shi Huang became the first king of a state that dominated all of China, something that was unprecedented. Eager to show that he was no longer a simple king, he created a title expressing the desire to unify the infinite territory of the Chinese kingdoms, in fact uniting the world (the ancient Chinese, like the ancient Romans, believed that their empire comprised the world as a whole).[4]

Qin Shi Huang refused to speak of death, never wrote a will, and in 212 B.C. he began to call himself the *Immortal*. Obsessed with immortality, he sent an expedition to the eastern islands (possibly Japan) in search of the elixir of immortality. The expedition never returned, supposedly for fear of the *Immortal* Emperor, as they had not found the desired elixir. Qin Shi Huang is believed to have died after drinking mercury, an element he hoped would make him immortal. He was buried in a large mausoleum with the famous terracotta warriors, more than 8000 soldiers with 520 horses. The mausoleum, near the present city of Xi'an, was discovered in 1974, although its burial chamber is still closed.

The elixir of immortality, a legendary potion that guaranteed eternal life, is a recurring theme in many cultures. It was one of the goals pursued by many alchemists, a remedy that cured all diseases (the universal panacea) and prolonged life eternally. Some alchemists, such as the Swiss doctor and astrologer Paracelsus, made great strides in the pharmaceutical field as a result of this quest. The magic elixir is related to the philosopher's stone, a mythical stone that would transform materials into gold and supposedly create that vital elixir.

Not only the ancient Egyptians and Chinese considered the possibility that there was an elixir of life. These ideas came or emerged independently in virtually every culture. For example, the Vedas groups in India also believed in a link between eternal life and gold. An idea they probably acquired from the Greeks after Alexander the Great's invasion of India in 325 B.C. Regarding these thoughts, it is also possible that from India it ended up in China, or vice versa. However, the idea of the elixir of life no longer has such an impact in India, since Hinduism, the country's first religion, professes other beliefs regarding immortality.

The fountain of youth is another of those legends that takes us back to the desire for eternity. Symbol of immortality and longevity, this legendary fountain would supposedly heal and restore youth to anyone who drank from its waters or bathed in them. The first known reference to the myth of a fountain of youth is in the third book of the *Histories* of Herodotus, from the fourth century BC. The *Gospel of John* narrates the episode of the Pool of Bethesda, in Jerusalem, where Jesus performs the miracle of healing a man apparently crippled. The oriental versions of *Romances of Alexander* tell the story of the "water of life" that Alexander the Great sought in the company of his servant. The servant in that story comes from the legends of Al-Khidr in the Middle East, a saga that also appears in the Koran. These versions were very popular in Spain during and after the Muslim era and would have been known by the explorers who traveled to America.

Native American stories about the healing source were related to the mythical island of Bimini, a country of wealth and prosperity located somewhere in the north, possibly in the location of the Bahamas. According to legend, the Spaniards learned about Bimini from the Arawak of Hispaniola, Cuba, and Puerto Rico. Bimini and its healing waters were then very popular topics in the Caribbean. The Spanish explorer Juan Ponce de León heard about the fountain of youth from Puerto Rico's natives when he conquered the island. Dissatisfied with his material wealth, he undertook an expedition in 1513 to locate it. He discovered the present state of Florida but never found the fountain of eternal youth.[5]

In today's so-called Western religions, based on monotheistic Abrahamic traditions such as Judaism, Christianity, Islam, and Baha'ism (Bahá'í Faith), for example, the path to immortality is achieved mainly through resurrection. On the other hand, in the so-called Eastern religions of today, based on the Vedic traditions of India such as Hinduism, Buddhism, and Jainism, the path to immortality occurs through reincarnation. Traditionally, in Western religions bodies have to be buried for resurrection, while in Eastern religions bodies have to be incinerated for reincarnation. But neither resurrection nor reincarnation are scientifically proven, and they are evidently part of the old mythological beliefs of pre-scientific times.

Israeli historian Yuval Noah Harari, of the Hebrew University of Jerusalem, has also studied the subject of immortality in depth in two of his major works: *Sapiens: A Brief History of Humankind*, originally published in 2011, and *Homo Deus: A Brief History of Tomorrow*, in 2016. The first book refers to the history of humanity from the beginning of the evolution of *Homo sapiens* to the political revolutions of the twenty-first century. Religions and the subject of death are a fundamental element in all these great historical events.

In his second book, Harari wonders what the world will be like in the years to come. We will be faced with a new set of challenges, and he tries to analyze how we will face them thanks to the enormous advances in science and technology. Harari explores the projects, dreams, and nightmares that will shape the twenty-first century, from overcoming death to the creation of artificial intelligence. Specifically, on the subject of immortality, Harari comments in the section "*The Last Days of Death*":[6]

> In the twenty-first century humans are likely to make a serious bid for immortality. Struggling against old age and death will merely carry on the time-honored fight against famine and disease, and manifest the supreme value of contemporary culture: the worth of human life. We are constantly reminded that human life is the most sacred thing in the universe. Everybody says this: teachers in schools, politicians in parliaments, lawyers in courts and actors on theatre stages. The Universal Declaration of Human Rights adopted by the UN after the Second World War – which is perhaps the closest thing we have to a global constitution – categorically states that 'the right to life' is humanity's most fundamental value. Since death clearly violates this right, death is a crime against humanity, and we ought to wage total war against it.
>
> Throughout history, religions and ideologies did not sanctify life itself. They always sanctified something above or beyond earthly existence, and were consequently quite tolerant of death. Indeed, some of them have been downright fond of the Grim Reaper. Because Christianity, Islam and Hinduism insisted that the meaning of our existence depended on our fate in the afterlife, they viewed death as a vital and positive part of the world. Humans died because God decreed it, and their moment of death was a sacred metaphysical experience exploding with meaning. When a human was about to breathe his last, this was the time to call priests, rabbis and shamans, to draw out the balance of life, and to embrace one's true role in the universe. Just try to imagine Christianity, Islam or Hinduism in a world without death – which is also a world without heaven, hell or reincarnation.
>
> Modern science and modern culture have an entirely different take on life and death. They don't think of death as a metaphysical mystery, and they certainly don't view death as the source of life's meaning. Rather, for modern people death is a technical problem that we can and should solve.

From Mythology to Science

In the last decades, impressive scientific advances have been made in all areas, including biology and medicine. In 1953 the structure of DNA was discovered, one of the most important advances in biology. This process has been accelerated by later discoveries such as embryonic stem cells and telomeres, for example. In medicine, the first heart transplant was performed in 1967, smallpox was eradicated in 1980, and great advances are now taking place in regenerative medicine, gene therapies such as CRISPR editing, therapeutic cloning, and organ bioprinting.

In the coming years, we will witness even greater and faster advances, thanks also to the widespread use of new sensors, the analysis of massive sets of data (known as "Big Data"), and the use of artificial intelligence for the improved interpretation and analysis of medical results. These advances are not occurring in a linear way, but exponentially. The speed with which the human genome was sequenced is a clear example of these exponential trends.

The Human Genome Project began in 1990, and by 1997 only 1% of the total genome had been sequenced. That is why some "experts" thought that we would need centuries to be able to sequence the remaining 99%. Fortunately, thanks to exponential technologies, the project concluded in 2003. As the American futurist Ray Kurzweil explains, every year since 1997 the sequenced percentage approximately doubled, i.e., 2% in 1998, 4% in 1999, 8% in 2000, 16% in 2001, 32% in 2002, 64% in 2003, and completed a few weeks later.[7]

Biology and medicine are being digitized rapidly, and this will allow exponential advances in the coming years. Artificial intelligence will help more and more, which will generate continuous positive feedback for further progress in all areas, including biology and medicine. On the other hand, experiments have already been initiated to extend life and rejuvenate different model animals, such as yeast, worms, mosquitoes, and mice.

Scientists in different parts of the world are already investigating how aging works and how to reverse it. From the United States to Japan, from China to India, through Germany and Russia. Groups of researchers also appear throughout Ibero-America, from Spain to Colombia, from Mexico to Argentina, passing through Portugal and Brazil. For example, a group of scientists under the direction of Spanish biologist María Blasco, director of CNIO (the Spanish National Cancer Research Centre) in Madrid, has created the so-called *Triple* mice, which live approximately 40% longer.[8] With totally different technologies, other scientists such as the Spanish Juan Carlos Izpisúa, an expert researcher at the Salk Institute for Biological Studies in La Jolla, California, have also been able to rejuvenate mice by 40%.[9] These types of experiments continue to progress, and it is likely that we will continue to increase longevity and rejuvenation in mice in the coming years.

Many other scientists from around the world, including several of the best international universities such as Cambridge, Harvard, MIT, Oxford, and Stanford, have groups interested in competing in the Methuselah Mouse Award, sponsored by the Methuselah Foundation in the United States.[10] An award has already been given to scientists who have managed to extend the life of mice to the equivalent of 180 human years,[11] but the full goal is to reach almost a thousand equivalent human years, like the legendary Methuselah of the *Old Testament*.

Experiments with mice have many advantages, as mice have relatively short lives (1 year in the natural state, 2 to 3 years in laboratory conditions) and their genomes have many similarities to the human genome (it is estimated that we share about 90% of the genome with mice). Scientists have experimented with different types of treatments and therapies, among which we can mention, for now, caloric restriction, telomerase injections, stem cell treatments, gene therapies, and more discoveries that we will continue to see in the coming years. This research is being done not because we love mice and want younger, longer-lived mice. The researchers, even though they may not say so publicly, look forward to implementing similar advances in humans to make us longer-lived and younger beings. Like many other people, sometimes scientists cannot say what they really think for fear of losing funding or other reasons, but the applications of this research are evident.

There are many scientists working with different types of model animals to stop and reverse aging. Two other examples of well-known North American scientists are Michael Rose of the University of California at Irvine, who has quadrupled the life expectancy of fruit flies *Drosophila melanogaster*,[12] and Robert J S Reis of the University of Arkansas for Medical Sciences, who has increased the longevity of nematode worms *C. elegans* by up to 10 times.[13] Again, the scientists' goal is not to get longer-lived flies and worms, but to use these discoveries to apply them to humans in due course.

Thanks to the important scientific advances of recent years, there are large and small companies that bet billions of dollars on scientific rejuvenation in humans. People are beginning to understand that this is a real possibility and closer and closer in time. The question today is not whether it will be possible, but rather when it will be possible. Therefore, billionaires like Peter Thiel, famous since PayPal, Jeff Bezos from Amazon, Sergey Brin and Larry Page from Alphabet/Google, Mark Zuckerberg from Facebook, Larry Ellison from Oracle, along with many others, and more and more, are investing in anti-aging biotechnology to reverse it. Google created Calico (California Life Company) in 2013 to "solve death,"[14] Microsoft announced in 2016 that it will cure cancer within 10 years,[15] and Mark Zuckerberg and his wife Priscilla Chan said they would donate virtually all their wealth to cure and to prevent all diseases in one generation.[16] Jeff Bezos started Altos Labs with other billionaires in 2021 to advance cell reprogramming technologies to allow rejuvenation treatments.[17] Altos Labs was actually promoted by Russian-born billionaire Yuri Milner, and there are other multimillion investments being made in Europe, the rich Arab countries, and now also in East Asia. In 2022, Saudi Arabia announced the creation of Hevolution (Health+Evolution) Foundation to finance at least $1 billion of research about longevity per year over the next two decades, and the United Arab Emirates are considering similar initiatives. We could add many other examples, and every day we will see more, because the advances do not stop.

Some of the world's best scientists are openly working on rejuvenation technologies. Just to give a widely known example, let's mention the case of the American geneticist, molecular engineer, and chemist George Church, professor of genetics at Harvard Medical School, professor of Health Sciences and Technology at Harvard and MIT, among many other positions, both academic and business (because it is necessary to take these ideas of life and death from academia to industry). Church, who was one of the pioneers of the human genome sequence and is considered a trailblazer of personal genomics and synthetic biology, has recently said:[18]

> Probably we'll see the first dog trials in the next year or two. If that works, human trials are another two years away, and eight years before they're done. Once you get a few going and succeeding it's a positive feedback loop.

The truth is that there is no scientific principle that prohibits rejuvenation and imposes the need for death. Not in biology, not in chemistry, not in physics. Consequently, the distinguished American physicist Richard Feynman, winner of the Nobel Prize in Physics, explained in a 1964 lecture entitled "The Role of Scientific Culture in Modern Society":[19]

> It is one of the most remarkable things that in all of the biological sciences there is no clue as to the necessity of death. If you say we want to make perpetual motion, we have discovered enough laws as we studied physics to see that it is either absolutely impossible or else the laws are wrong. But there is nothing in biology yet found that indicates the inevitability of death. This suggests to me that it is not at all inevitable, and that it is only a matter of time before the biologists discover what it is that is causing us the trouble and that that terrible universal disease or temporariness of the human's body will be cured.

In recent years a number of scientific publications have been created on advances in new areas such as the study of rejuvenation and anti-aging. One of them is *Aging* magazine, which published its first issue in 2009, when three of its editors, Russian-American scientist Mikhail V. Blagosklonn, American

Judith Campisi, and Australian David A. Sinclair, wrote the inaugural article entitled "Aging: Past, Present and Future":[20]

> In his *Foundation series*, published in the 1950s, Isaac Asimov imagined Civilization capable of colonizing the entire Universe. This feat is unlikely to occur. Strikingly, Asimov referred to a 70-old man as an old individual who is unlikely to live much longer. Thus, in literature's most daring fantasy, the pace of aging could not be slowed. Yet, given the present pace of discovery in the aging field, this feat might become a reality within our life time, with science surpassing science fiction.
>
> PAST
>
> Once August Weismann had divided life into a perishable soma and immortal germ line, the soma began to be viewed as disposable. As Weismann wrote in 1889, "the perishable and vulnerable nature of the soma was the reason why nature made no effort to endow this part of the individual with a life of unlimited length".
>
> PRESENT
>
> The first successful screens for genes that postponed aging began in the mid 1980's. Despite common opinion that genes that control aging were unlikely to exist, Klass performed a mutagenesis screen for longlived *C. elegans* mutants and found candidates, one of which, age-1, was characterized by Johnson and colleagues. In 1993, Kenyon and colleagues, also screening for long-lived *C. elegans*, found that mutations in the gene daf-2 increases the longevity of *C. elegans* hermaphrodites by more than two-fold compared to wild type nematodes. Daf-2 was already known to regulate formation of the dauer state, a developmentally arrested larval form that is induced by crowding and starvation. Kenyon et al. suggested that the longevity of the dauer results from a regulated life span extension mechanism. This discovery provided entry points into understanding how life span can be extended.

The editors take a quick look at the incipient discipline of the study of aging, explaining the scientific beginnings at the end of the nineteenth century and the great advances throughout the twentieth century, especially in the last two decades. In fact, it was only in the 1980s that genes directly related to cell aging were found in small nematode worms called *C. elegans*. Since then, there has been a greater and better understanding of the aging process, how it occurs, and even how to reverse it.

However, the fact that proof for this concept already exists does not mean that we know how to do it. In fact, we do not yet know, and that is why numerous experiments are being carried out with different therapies and treatments, and on different types of organisms, to find out what works and why. It's not easy at all, and it probably won't be easy, but we do know it's possible. In fact, the question is no longer whether it is possible, but when it will be possible to develop and commercialize the first scientific treatments to rejuvenate human beings. We are not worms, nor are we mice, so many of the things we discover with worms or mice will probably not be directly applicable to humans. However, they will point out several possibilities, which thanks to advances such as the use of Big Data and artificial intelligence, among others, will help us move faster to find possible cures to human aging.

Blagosklonn, Campisi, and Sinclair start from the past and the present to also indicate what might happen in the future, along with some of the possible treatments and therapies for aging and age-related diseases. For now, and in the case of this introductory book aimed at a general audience, it is not necessary to know in depth the details (or acronyms such as DNA, AMPK, RNA, FOXO, IGF, mTOR, NAD, PI-3K, CR, TOR, and many others even more convoluted). However, we mention it here as a general summary of the great discoveries of today and tomorrow, as the authors indicate in their article:

FUTURE

Of great interest and excitement, aging now appears to be regulated at least in part by signal-transduction pathways that can be manipulated pharmacologically. Prototype anti-aging drugs are available now to treat age-related diseases, and are predicted to slow aging processes. Modulators of sirtuins have been discovered that mimic CR and mitigate certain age-related diseases. The TOR pathway is another target. Ironically, TOR itself was discovered as target of rapamycin in yeast (Sirolimus or Rapamune), a clinically available drug that is tolerated even when taken in high doses for several years. Rapamycin has potential as a therapy for most of not all age-related diseases, and metformin, an anti-diabetic drug and activator of AMPK, which acts in the TOR pathway retards aging and extends lifespan in mice.

Thus, recent paradigm shifts in aging research has put signaling pathways (growth-promoting pathways, DNA damage responses, sirtuins) at front stage, has established that aging can be regulated and can be inhibited pharmacologically.

At this opportune time, *Aging* (Impact Journal on Aging or Impact Aging) is launched. This journal embraces the new gerontology. Recent breakthrough in gerontology is due to an integration of different disciplines, such as genetics and development in model organisms, signal transduction and cell cycle control, cancer cell biology and DNA damage responses, pharmacology, and the pathogenesis of many age-related diseases. The journal will focus on signal transduction pathways (IGF- and insulin-activated, mitogen-activated and stress-activated pathways, DNA damage response, FOXO, Sirtuins, PI-3K, AMPK, mTOR) in health and disease. Topics include cellular and molecular biology, cell metabolism, cellular senescence, autophagy, oncogenes and tumor suppressor genes, carcinogenesis, stem cells, pharmacology and anti-aging agents, animal models, and of course age-related diseases such as cancer, Parkinson's disease, type II diabetes, atherosclerosis, macular-degeneration, which are deadly manifestations of aging. The journal will also embrace articles that address both the possibilities and the limits of the new science of aging. Of course, the possibility that the diseases of aging can be delayed or treated by drugs that affect the overall aging process, thus potentially extending healthy lifespan, is long-standing dream of mankind.

When this visionary article was published in 2009, almost nothing was known about one of today's most powerful genetic technologies: the famous CRISPR (discovered at the end of the 1980s, its first applications only began to be developed at the beginning of the decade of 2010). The human genome sequence officially ended in 2003, and the cloned sheep Dolly was born in 2006. The first induced pluripotent cells (usually abbreviated as iPS cells) were obtained in 2006, but the first treatments did not appear until the decade of 2010. The *Aging* journal witnessed huge transformations in less than a decade since it was first published in 2009 and will witness even more changes over the next decade. All this must be brought into perspective to understand the frenetic pace of progress, as the same will happen in the next 10 years, or perhaps just 4 or 5 years, since progress continues to accelerate. We are convinced that in 2 or 3 years we will see progress so impressive that we will have to rewrite several parts of this book.

Another excellent journal on these topics is *Rejuvenation Research*, launched in 1998 and later under the direction of British biogerontologist Aubrey de Grey. In the two decades since its inception, the journal has witnessed great advances reported in its articles, and we hope that these will continue to advance exponentially in the coming years.[21]

The *Appendix* to this book shows a comprehensive chronology that enables us to contextualize the rapid advances in our understanding of life on Earth. This chronology, moreover, attempts to predict some of the fascinating possibilities that may take place in the coming decades thanks to the exponential changes that are approaching, in the opinion of the authors, as well as further references from experts such as the famous aforementioned American futurist Ray Kurzweil.

From Science to Ethics

We have already discussed how we have succeeded in extending the life of worms and mice, among other animals. Why do we experiment with them? Are scientists looking for younger, longer-lived worms and mice? Of course not. One of the goals is to understand how aging and rejuvenation work in order to start clinical trials in humans at some time in the future, as we have already commented and as we will continue to insist throughout these pages.

If we now accept that it will be possible to extend human life thanks to future scientific advances, then we must discuss whether it will also be ethical. Our answer is that it is not only ethical but also our moral responsibility. However, there are still some very influential people (the so-called *influencers*), such as American businessman and philanthropist Bill Gates, who do not seem convinced of the priority of curing aging. When asked at a public event on the Reddit website what he thought of life-extending and immortality research, Gates answered:[22]

> It seems pretty egocentric while we still have malaria and TB for rich people to fund things so they can live longer. It would be nice to live longer though I admit.

However, the same critique could be made against numerous medical research programs, such as those underway to cure cancer or heart disease. Repairing these diseases will also extend life. While people still die of malaria and tuberculosis, diseases that can be treated at relatively low cost, it might seem a misplaced priority to invest large sums in searching for cures for cancer and heart disease. If the criterion is really to save as many lives as possible by spending a certain amount of money, then we might ask ourselves: wouldn't it be better to cancel cancer research initiatives and instead buy more mosquito nets and ensure that they are distributed to all areas still suffering from malaria? Obviously not, which shows that things are far from being black and white.

Actually, the main cause of death on the planet is neither malaria nor tuberculosis: it is aging. A successful rejuvenation project would therefore meet all of the above requirements. Pursuing that goal is far from being egocentric or narcissistic. Not only researchers (and their loved ones) will benefit from the program. The benefits can reach the entire planet, including citizens of the poorest communities still suffering from outbreaks of malaria and tuberculosis. After all, these communities are also suffering from aging.

The greatest cause of suffering in the world is aging and the age-related diseases that lead to death. Today, about 150,000 people in the world die every day.[23] Two thirds of these deaths are due to age-related diseases. In more advanced countries, the number is much higher, with nearly 90% of people dying due to aging and related major diseases such as neurodegenerative, cardiovascular, or cancer.

Aging is a tragedy difficult to compare with any other. More people die in the world every day due to aging than all the other causes of death put together. Specifically, over two times as many die from aging as from all other causes, including malaria, AIDS, tuberculosis, accidents, wars, terrorism, famines, etc. Aubrey de Grey explains it in a very clear and direct way:[24]

> Aging really is barbaric. It shouldn't be allowed. I don't need an ethical argument. I don't need any argument. It's visceral. To let people die is bad. I work to cure aging, and I think you should too, because I feel that saving lives is the most valuable thing anyone can spend their time doing, and since over 100,000 people die every single day of causes that young people essentially never die of, you'll save more lives by helping to cure aging than in any other way.

The great enemy of humanity is death caused by aging. Death has always been our worst enemy. Fortunately, deaths from wars and famines have decreased considerably today, in addition to past infectious diseases such as polio and smallpox. The main common enemy of all humanity is not religions, diversity of ethnic groups, different cultures, wars, terrorism, ecological problems, environmental pollution, earthquakes, distribution of water or food, etc. Without denying the suffering that these other factors can cause, the greatest enemy of humanity in our time is, by far, aging – and diseases related to aging.[25]

The suffering caused by aging to each individual, their families, and society as a whole is difficult to quantify, but we emphasize that it is far greater than any other current tragedy. Life is considered "sacred" by most religions, and it is the first right of people, since without life there is no other right or duty that is worthwhile. The right to life is a right that is granted to anyone, a right that protects from being deprived of life by others. This right is generally recognized by the simple fact of being alive, is considered a fundamental right of every person, and is included not only in the *Universal Declaration of Human Rights* but also explicitly in the overwhelming majority of advanced legislations.

Legally, among the rights of man, the most important is undoubtedly the right to life, since it is the reason for the existence of others, since it would not make sense to guarantee property, religion, or culture if the subject to whom it is granted is dead. It falls within the category of Civil and First-generation Rights and is recognized in numerous international treaties: the *International Covenant on Civil and Political Rights,* the *Convention on the Rights of the Child,* the *Convention on the Prevention and Punishment of the Crime of Genocide,* the *International Convention on the Elimination of All Forms of Racial Discrimination,* and the *Convention against Torture and Other Cruel, Inhuman or Degrading Treatment or Punishment.* The right to life is clearly enshrined in the third article of the *Universal Declaration of Human Rights*:[26]

> Everyone has the right to life, liberty and security of person.

In a striking piece of writing, *The Fable of the Dragon-Tyrant* compares human aging with a tyrant dragon that devours thousands of lives each day. Our social system has adapted to this fatality by investing huge amounts of money and adapting our psychology to this enormous tragedy. The fable was originally written in 2005 by philosopher Nick Bostrom, director of the *Future of Humanity Institute,* part of the Faculty of Philosophy at the University of Oxford, and co-founder of the *World Transhumanist Association* (now known as *Humanity*+).[27] It was turned into a captivating short video in 2018 by YouTuber CGP Grey.[28]

A Revolution for Everyone: From Children to the Elderly

As we have already advanced and will discuss in the next chapters, the scientific possibility of physical immortality and its moral defense is actually humanity's greatest challenge. It has always been the most desired dream, since the appearance of the first *Homo sapiens sapiens*, but we never had the technology to realize that immortal dream, until today.

Even children understand that aging is bad, that death is the most horrible loss that can happen to someone and their family. The Belarusian-American writer Gennady Stolyarov II, head of the Transhumanist Party in the United States, wrote a children's book in 2013 under the title *Death is Wrong*, where he explains:[29]

> This is the book I would have wanted to have as a child, but did not. Now that you have it, you can discover in less than an hour what it took me years to learn

in bits and pieces. You can instead spend those years fighting the greatest enemy of us all: death.

Stolyarov continues a conversation he held as a child with his mother, who was explaining to him that people finally "die." The young boy asks his mother in surprise:

"Die? What does that mean?" – I asked – "It means they stop existing. They are just not there anymore" – she replied –
   "But why do they die? Do they do anything bad to deserve it?" – I questioned – "No, it happens to everyone. People get old and then die" – she said – "It is wrong!" – I exclaimed – "People should not die!"

Fortunately, the children of the present generation can belong to the first generation of immortal (or amortal) humans. If we continue to advance exponentially, we may soon have the first treatments and therapies for human rejuvenation. And the sooner, the better. As American actress, singer, comedian, scriptwriter, and playwright Mae West said: "You are never too old to become younger!"

We must be aware that we live between the last mortal generation and the first immortal: where do you want to be? No matter how old you are now, we recommend that you join this revolution against aging and death. As it is written in 1 Corinthians 15:26 of the *Bible*:

The last enemy to be destroyed is death.

Notes

1. https://www.amazon.com/Immortality-Quest-Forever-Drives-Civilization/dp/1510716157
2. https://www.amazon.com/Egyptian-Book-Dead-Integrated-Full-Color/dp/1452144389/
3. https://www.amazon.com/Epic-Gilgamesh/dp/014044100X
4. https://www.amazon.com/First-Emperor-China-Jonathan-Clements-ebook/dp/B00XJIQ7K2/
5. https://www.amazon.com/EUROPEAN-DISCOVERY-AMERICA-D-1492-1616/dp/B000J57YR8
6. http://www.openthemagazine.com/article/essay/the-last-days-of-death
7. https://www.youtube.com/watch?v=h6tYxQnxRj8

8. http://www.encuentroseleusinos.com/work/maria-blasco-directora-del-cnio-envejecer-es-nada-natural/
9. https://elpais.com/elpais/2016/12/15/ciencia/1481817633_464624.html
10. https://www.mfoundation.org/
11. https://web.archive.org/web/20190324131618/http://www.sens.org/outreach/conferences/methuselah-mouse-prize
12. https://www.faculty.uci.edu/profile.cfm?faculty_id=5261
13. https://medicine.uams.edu/biochemistry/faculty/secondary/reis/
14. http://time.com/574/google-vs-death/
15. http://www.telegraph.co.uk/science/2016/09/20/microsoft-will-solve-cancer-within-10-years-by-reprogramming-dis/
16. http://www.businessinsider.com/mark-zuckerberg-cure-all-disease-explained-2016-9
17. https://www.technologyreview.com/2021/09/04/1034364/altos-labs-silicon-valleys-jeff-bezos-milner-bet-living-forever/
18. https://www.fightaging.org/archives/2017/12/george-church-discusses-gene-therapies-as-a-basis-for-therapies-to-control-aging/
19. http://www.amazon.com/Pleasure-Finding-Things-Out-Richard/dp/0465023959
20. https://dash.harvard.edu/bitstream/handle/1/4931360/2815757.pdf
21. https://home.liebertpub.com/publications/rejuvenation-research/127
22. https://www.reddit.com/r/IAmA/comments/2tzjp7/comment/co3q1lf/
23. https://www.theguardian.com/technology/2010/aug/01/aubrey-de-grey-aging-research
24. https://www.fightaging.org/archives/2004/11/strategies-for-engineered-negligible-senescence/
25. https://www.amazon.com/Advancing-Conversations-Advocate-Indefinite-Lifespan/dp/1785353969
26. https://www.un.org/en/about-us/universal-declaration-of-human-rights
27. https://nickbostrom.com/fable/dragon.html
28. https://www.youtube.com/watch?v=cZYNADOHhVY
29. http://www.rationalargumentator.com/index/death-is-wrong/

Testimonials

Enter any drugstore or bookstore, and we confronted with a mountain of nonsense concerning the aging process. Society seems obsessed with aging. That is why *The Death of Death* is such a refreshing delight, able to cut through the hype and reveal a balanced, authoritative, and lucid discussion of this controversial topic. It summarizes the astonishing breakthroughs made recently in revealing how science may one day conquer the aging process.
Michio Kaku, theoretical physicist and author of *The God Equation: The Quest for a Theory of Everything*

We are entering a *Fantastic Voyage* into life extension, crossing different bridges that will take us to indefinite life spans. *The Death of Death* explains clearly how we might soon reach longevity escape velocity and live long enough to live forever.
Ray Kurzweil, co-author of *Fantastic Voyage: Live Long Enough to Live Forever* and co-founder of Singularity University

José and David have captured the spirit of what may be the greatest revolution in history. Not some abstract far distant promise but an exponential growth of rigorous discovery and technology in our midst but understood by only a few. This clear prose will help you join the conversation and choose a path. You should definitely read this international bestseller!
George Church, professor at Harvard University and Massachusetts Institute of Technology, and founding member of the Wyss Institute for Biologically Inspired Engineering

The Death of Death is a truly revolutionary book. This is a visionary book that confronts us with the terrible reality of aging, and its authors are friends and connoisseurs of the subject. I believe that the authoritative and exhaustive description of this crusade that José and David make in this excellent book will accelerate this process. Forward!
Aubrey de Grey, founder of LEV (Longevity Escape Velocity) Foundation and co-author of *Ending Aging*

This wonderful book presents a compelling case for truly unprecedented life extension. *The Death of Death* shows how the time has come at last for Death to get a taste of its own medicine!
Terry Grossman, co-author of *Fantastic Voyage: Live Long Enough to Live Forever* and longevity expert

It is a fact that the science of longevity is now catching up with the aspiration that most people have to live longer lives in good health. Very soon this will become a reality and this is brilliantly explained in the new book *The Death of Death*: a must read for anyone interested in the evolving science of Juvenescence.
Jim Mellon, co-author of *Rejuvenescence* and multimillionaire investor in longevity research

The Death of Death is a critical book for anyone who is alive and would prefer to remain that way. It gives a clear, rich and entertaining overview of the near-term prospects for using technology to eliminate disease and death. I expect *The Death of Death* to make major headway in furthering the world's understanding of both HOW technology is going to eliminate death, and WHY this is a tremendously good thing.
Ben Goertzel, chair of HumanityPlus and CEO of SingularityNET

The Death of Death offers a wonderful vision of how the breathtaking scientific advances in aging research and other fields can abolish human death by scientific means and why this is the right path for humankind.
João Pedro de Magalhães, chair of Molecular Biology at the University of Birmingham

The Death of Death is a very interesting book, because it is likely that the following generations live, on average, much more than they expect. Technology advances by leaps and bounds (unless you are waiting for an

immediate cure), and therefore it is important to discuss what to do with much more, and more, life. Read *The Death of Death* and think...
Juan Enriquez, author of *As the Future Catches You* and *Evolving Ourselves*

The authors are two of the main lights of the world futuristic movement. *The Death of Death* considers the technologies and ethical issues that will make this book a reference to human longevity for educators and politicians.
Martine Rothblatt, founder of United Therapeutics and author of *Virtually Human*

I do not have the scientific competence to know if what is explained in *The Death of Death* is viable, but I do know that we will hardly find in our days provocateurs, dreamers and agitators of consciences more fearsome than these authors. This excellent book, which explains complex issues in a simple way, is the best proof.
Álvaro Vargas Llosa, international policy expert and co-author of the *Guide of the Perfect Latin American Idiot*

We have seen incredible advances in our understanding of aging in recent years. What will we see in the coming years? Will we be able to cure aging? And if so, when? If you want to know, *The Death of Death* is your book.
Zoltan Istvan, former US presidential candidate and author of *The Transhumanist Wager*

Reading this book is the best way to start a new day full of life and opportunities. If rejuvenation occurs in the short term, we must begin now to consider this extension of life that the authors offer us in *The Death of Death*.
Diego Arria, former president of the United Nations Security Council

The Death of Death deals with one of the top moral priorities today: to slow and stop aging and death. As the case for the scientific feasibility of this becomes clearer, the importance of explaining and understanding the implications grows ever larger. This book fills that important role.
Anders Sandberg, professor of the Future of Humanity Institute, University of Oxford

I enthusiastically recommend *The Death of Death* to our followers and also to skeptics who do not realize how close science is to eradicating biological aging forever.
Bill Faloon, co-founder of the Life Extension Foundation and author of *Pharmocracy*

Lengthening telomeres could be one of the keys to cure aging and stop the decline in health. Read *The Death of Death* and discover how we can stop and reverse aging.

Bill Andrews, founder of Sierra Sciences and co-author of *Curing Aging*

As medicine becomes an information technology, genetic and cellular therapies allow us to manipulate cells with greater precision. Thanks to the exponential acceleration of these capacities, aging and, thereby, death will be eliminated. *The Death of Death* explains the details of this campaign, and how humanity will soon live long enough to live forever.

Elizabeth Parrish, founder of BioViva and "patient zero" in telomerase treatments

Aging is the main cause of almost all diseases and death, so accelerating progress in the biotechnology of longevity is the most altruistic cause anyone can pursue. We live in the most exciting moment in the history of humanity, since we can substantially increase the productive longevity in our lifetimes. *The Death of Death* presents the many moral and economic arguments to eliminate death as we know it and analyzes recent trends in science that can bring us closer to this goal. Read this book and join the revolution against aging: the true emperor of all diseases.

Alex Zhavoronkov, founder and CEO of Insilico Medicine and author of *The Ageless Generation*

The Death of Death is a visionary book. Something very important if we are going to win the war against death, and every battle requires a meaning, a "why," and this book serves as inspiration.

Jason Silva, futurist and presenter of *Mind Games* with National Geographic

We are headed for a future where death shall have no dominion. Almost no one in the world understands this, but the authors get it, deeply, and in their excellent book *The Death of Death* they offer to take us by the hand, and show us the way forward, along the many paths technological and personal, to a promise of longer and healthier life.

Raymond McCauley, scientist, entrepreneur and founding faculty at Singularity University

The Death of Death brings together important research and insights to a future where we live much longer that we think is possible today. Can we really become immortal? If so, when?
Jerome Glenn, CEO of The Millennium Project

José and David bring passion informed by their deep knowledge in a field of human flourishing that is mostly unexplored today, but that is going to be one of features fundamentally shaping civilization tomorrow. If you are planning to live in the future, and you will, then you must read this book!
David Orban, investor and founder of the Network Society Ventures

The Death of Death is at once a description of our evolving understanding of the possibilities of immortality, and also a manifesto for their achievement. The authors, José Cordeiro and David Wood, are pioneering visionaries, but also ardent activists, for the steps we need to follow, as societies, to achieve the great rewards that await with extended and potentially immortal life.
Kenneth Scott, managing investor for Human Longevity and Rejuvenation Syndicate at VivaSparkle.com

This book is written with clarity and wisdom in telling the most important story of all time: *The Death of Death*. The prevention of death must become a global priority. Above all other existential risks, death is the most immediate threat to our species. This realization is so powerful that one day in the future the death of a single person will set off alarms throughout the entire world.
Natasha Vita-More, author of *Persistence of Long-Term Memory* and executive director of HumanityPlus

The Death of Death is a call to action. If new longevity technologies can be developed faster, many lives will be saved. It is an important and humanitarian message that everyone should listen to.
Sonia Arrison, associate co-founder of Singularity University and author of *100 Plus*

Humanity is headed towards an awakening of historic proportions as two worlds collide: exponential medical innovations and the archaic current healthcare establishment. *The Death of Death* is a crucial literary work that prepares readers for this exciting point of convergence. Understanding that a pivotal change in lifespans and healthspans is "ad portas" allows us to shape a future we want to be a part of.
Sergio Ruiz, co-founder of Turn Biotechnologies and Methuselah Fund

For the first time in human history, the fountain of youth might be within reach. José and David's optimistically – and perhaps prophetically – titled book, *The Death of Death*, is by far the finest discussion of this contemporary quest, what it could mean for each of us – and for humanity. A must-read for anyone who might one day die.

Robert C. Wolcott, co-founder and chair of TWIN Global

Sigmund Freud once said: "the goal of all life is death." I say: "the goal of all death is life." And in *The Death of Death*, José and David prepare us for this transition.

Sergey Young, founder of Longevity Vision Fund and author of *Growing Young*

The Death of Death is a bold, fascinating and engrossing book presenting the concept of longevity escape velocity that once achieved will allow us to stay young and healthy indefinitely. I couldn't put it down. Scientific discoveries of the last two decades brought a clear understanding of human aging and now the best minds in the world are working to cure aging. I have full confidence that our scientists will be successful, since I myself have participated in the amazing leaps of the computer and networking evolution and now the same thing is happening in longevity biotech. Proactive rejuvenation medicine will be the biggest market ever, and now is the best time to invest.

Fiona Miller, managing partner at quadraScope Venture Fund

In the age of digital health technologies, artificial intelligence and the rise of longevity research, it's particularly important to discuss issues around death with style and enough compassion. *The Death of Death* bravely examines the topic nobody wants to talk about and analyzes a future in which death itself might become obsolete.

Bertalan Meskó, founder of The Medical Futurist

I have known José and David for almost a decade, and can think only of a few people who have dedicated so much work, passion, and energy towards the techno-ethical humanitarian mission of Longevity Industrialization than them. José and David, through their years of international community building, breaking barriers and building bridges for the global longevity community, have made an incalculable impact towards bringing this all-important topic to the attention of new eyes and ears, helping to wake humanity up and recognize the most important inflexion point of its entire history (and future).

Dmitry Kaminskiy, general partner of Deep Knowledge Group and author of *Longevity Industry*

The field of longevity medicine is utterly multidisciplinary, and collaborative effort with various arenas, including biotech, AI, computational science, biology and geroscience are determining the speed of clinical translation towards longevity diagnostics and interventions. Longevity medicine tackles risks of developing age-related comorbidities, and while it's not a priority to extend the lifespan, but the healthspan, it is important to know the future trends presented in *The Death of Death*.

Evelyne Bischof, professor at Shanghai University of Medicine and Health Sciences and visiting professor at Tel Aviv University School of Medicine

José and David are real disruptors and pioneers in the field of life-extension and longevity. Their provocative and visionary book gives a good overview about the massive changes which are ahead of us affecting all aspects of our lives.

Marc P. Bernegger, founding partner of Maximon

This book X-rays the possibilities of using advanced science and technology to access quality lifespan and overcome death caused by aging. As a result, ethical application and putting humanity at the center of it all should be key. *The Death of Death* is a book worth reading.

Brenda Ramokopelwa, CEO of Transdisciplinary Agora for Future Discussions and director at AfroLongevity

Our body is the greatest asset, but it is often poorly maintained. Implementation of geroscience diagnostics and interventions is revolutionizing health(care) and will optimize healthspan. How long this span of health should last is rigorously discussed in *The Death of Death*. Life or health extension?

Andrea B. Maier, professor in Medicine and Healthy Aging and co-director Centre for Healthy Longevity, National University of Singapore

Aging research has become mainstream and, while most of the current focus is understandably on extending healthy longevity and offsetting the emerging demographic, global crisis of age-related chronic disease, the future impact of longevity research is an equally important and fascinating issue. No one is a bigger advocate of the future of this field than José and his coauthor David. Here, they thoughtfully explore whether age-associated mortality is truly inevitable. For the first time in history, this is a legitimate question with which scientific endeavour must grapple.

Brian Kennedy, director of Centre for Healthy Longevity and professor of Biochemistry and Physiology, National University of Singapore

Death may not yet be dead, but we humans are fast-developing the tools to put up, quite literally, the fight of our lives. *The Death of Death* thoughtfully explores a future of human mortality that could be radically different from our past. It should be required reading for anyone interested in the promising new science of human life extension.

Jamie Metzl, founder of OneShared.World and author of *Hacking Darwin: Genetic Engineering and the Future of Humanity*

The Death of Death gives us a vision of a world where death needs not be a given. No one should dismiss the importance of this book or the logic with which co-authors José and David construct it. They are two of the world's leading futurists on aging. Open your mind and heart and leap into a future that may already be at the tipping point of human change and destiny. What would extreme longevity or *The Death of Death* mean to government policies and culture? Would you live your life differently? As the book grapples with these issues, the authors provide answers not just for immortalists but for all of us.

Adriane Berg, host of Generation Bold Podcasts, fellow of *The New York Times Age Boom* and co-founder of National Academy of Elder Law Attorneys

There are so many questions about our relationship with death, many of them unspoken; there is a dissonance in our group think – from individuals, to politicians, to scientists. *The Death of Death* not only deals with these questions head-on, it sets-out the technical and moral roadmap for humankind to embrace longevity and for individuals to plan to live as long as they wish. Longevity is the next megatrend and this book makes a huge contribution to its foundation.

Phil Newman, editor-in-chief of *Longevity.Technology*

The Death of Death is a masterful exploration of the exponential developments in science and technology enabling us to unravel the intricacies of the aging process and imagine the tools to delay and maybe even stop death. We need brilliant minds and provocateurs like José and David to challenge pervasive thinking that constrains progress to live longer healthier. The next frontier in longevity is how we can harness and share collective intelligence across all of society to create a world where all people can flourish – since hope, passion and curiosity and all the unique magical ingredients of the human spirit that drive us are critical to give people the longing to live, and not just survive.

Tina Woods, founder & CEO of Collider Health, Healthy Longevity Champion, National Innovation Centre for Aging

I am deeply grateful to both José and David writing this powerful book. It not only covers the rapidly expanding science of curing aging and death, but reminds us that we as human beings have historically always sought out immortality, because it's innate in our bodies to ultimately conquer death.
James R. Strole, director for the Coalition for Radical Life Extension and RAAD (Revolution Against Aging and Death) Fest

The fight against aging should be something uniting us all. *The Death of Death* delivers an important, thorough and enjoyable deep dive into how we could win this fight and what will come next.
Nicklas Brendborg, associate of Futures Institute AG and author of *Jellyfish Age Backwards*

The fear of death has fuelled the quest by humans to extend their lives life since time immemorial. Although spiritual lives may last for eternity, recent advances in science and technology brings mankind a step closer towards realistically extending their mortal life through increasing health-span. This provides a foundation for many of the concepts persuasively covered in *The Death of Death* to inspire the next generation of scientists, entrepreneurs, educators, and policymakers to collectively endeavour to realise some of them over the next few decades.
Richard Siow, director of Aging Research, King's College London

The Death of Death is a revolutionary book that considers the new possibilities of "moonshot" thinking applied to longevity and rejuvenation technologies. Indeed, immortality is impossible until it becomes possible, this is disruptive thinking!
Naveen Jain, founder of Viome and Moon Express, vice chair of Singularity University, author of *Moonshots*

We see an acceleration of scientific discoveries and clinical trials, clearly showing that we can slow down and reverse the aging process. The major problem now is the general perception of this effort as something "unnatural" or "impossible". *The Death of Death* is an excellent elaboration and argumentation to support our vision and general awareness. It will not leave anyone in doubt after reading.
Petr Sramek, managing general partner of LongevityTech.Fund

The Death of Death sets a new standard for living well for as long as we want. This book presents a logical roadmap for eternal life. Thrilling! Inspiring!! Logical!! Backed by very credible data from the top experts in the field.
John Asher, CEO and co-founder of Asher Longevity Institute

Contents

About the Authors

José Cordeiro, PhD José is an international fellow of the World Academy of Art and Science, vice chair of HumanityPlus, director of The Millennium Project, founding faculty at Singularity University in NASA Research Park, Silicon Valley, and former director of the Club of Rome (Venezuela Chapter), the World Transhumanist Association and the Extropy Institute. He has also been invited faculty at the Institute of Developing Economies IDE – JETRO in Tokyo, Japan, the Moscow Institute of Physics and Technology (MIPT), and the Higher School of Economics (HSE) in Russia.

José studied engineering at the Massachusetts Institute of Technology (MIT) in Cambridge, MA, economics at Georgetown University in Washington, DC, management at INSEAD in Fontainebleau, France, and science at Universidad Simon Bolivar in Caracas, Venezuela. He is a leading expert on technological change and future trends. He has published more than 10 books in 5 languages and appeared in programs with the BBC, CNN, Discovery Channel, and History Channel, among many other international media interviews.

He is a lifetime member of the Sigma Xi (ΣΞ) and Tau Beta Pi (ΤΒΠ) science and engineering honor societies and has received several prizes, including the Spanish Health Award by Instituto Europeo for promoting research on longevity and life extension. He has been Spanish candidate to the European Parliament in 2019, when he proposed the creation of the European Anti-Aging Agency.

David Wood, ScD David was a pioneer of the smartphone industry, co-founding in 1998 Symbian, the creator of the world's first successful smartphone operating system. Software written by his teams was included in half a billion smartphones over the following years, from companies such as Nokia, Motorola, Sony Ericsson, Samsung, LG, Fujitsu, and Panasonic.

David also spent 3 years as CTO of Accenture Mobility. While at Accenture, he co-led the company's international Mobility Health business initiative. He is now a full-time futurist speaker, analyst, and writer. He is the author or lead editor of several books, including *Anticipating 2025*, *Smartphones and Beyond*, *The Abolition of Aging*, *Sustainable Superabundance*, *Vital Foresight*, and, most recently, *The Singularity Principles*.

He also heads up London Futurists, a non-profit networking meetup with approaching 10,000 members, and has chaired over 250 public events on technoprogressive and futurist topics. In 2009 he was included in T3's list of "100 most influential people in technology." He has an MA in Mathematics from Cambridge University and an honorary doctorate in science from the University of Westminster. He is a Director of the IEET (Institute for Ethics and Emerging Technology) and the LEV (Longevity Escape Velocity) Foundation and was formerly Secretary of the international HumanityPlus organization.

1

Life Arose to Live

All men by nature desire to know.
Aristotle, circa 350 BC

Thy life's a miracle.
William Shakespeare, 1608

All truths are easy to understand once they are discovered; the point is to discover them.
Galileo Galilei, 1632

The world has come a long way since the first historical narratives about the creation of the universe proposed by primitive cultures. We have gone from pre-scientific mythological stories to scientific theories that can be assessed on the basis of experimentation. In any case, the origin of life is still a mystery that we hope will eventually be better understood.[1]

In 1924, the Russian scientist Aleksandr Oparin put forward his first ideas in his work *The Origin of Life on Earth*. Oparin was a convinced evolutionist, and therefore outlined a sequence of events by which these first organic substances were gradually transformed by natural selection to form a living organism in the primitive sea of the Earth.

Years later, in 1952, the young Stanley Miller, a chemistry student at the University of Chicago, together with his professor Harold Urey, tried to test this theory with a simple device that mixed water vapor, methane, ammonia and hydrogen. These gases were thought to be the gases that existed in the Earth's atmosphere at that time. Electrodes were used to simulate the electrical currents of primeval storms (energy inputs). With this experiment they simulated the prebiotic conditions and, thanks to the energy contribution made by the electrodes, they were able to obtain amino acids, some sugars and nucleic acids, but they never obtained living matter, only some of its components.

J. Cordeiro, D. Wood, *The Death of Death*, Copernicus Books,
https://doi.org/10.1007/978-3-031-28927-9_1

In 1953, English scientists Francis Crick and Rosalind Franklin and American James Watson discovered the structure of DNA. This discovery would forever mark later works and theories about the origin of life. The Spanish scientist Joan Oró then tried to merge advances in chemistry with the growing importance of DNA studies, following the 1955 advances of his compatriot Severo Ochoa. In 1959 he managed to synthesize adenine (one of the bases of DNA and RNA) in conditions that were believed to exist in the primitive Earth. In his book *The Origin of Life*, Oró wrote:[2]

> Some of the prebiotic processes are broadly reproducible in the laboratory, and the aqueous or liquid medium has been found to be the most suitable for their development. It is therefore almost certain that life came from what has been called the primitive ocean.

Bacteria Colonize the World

Regardless of how life originated on the planet, and perhaps we will never know, the truth is that the first living organisms must have been very small, simple cells with the capacity to multiply. These primitive microorganisms were probably bacteria, or something similar to the simplest bacteria we know today.[3]

Bacteria are the most abundant organisms on the planet. They are ubiquitous, found in all terrestrial and aquatic environments; they grow even in the most extreme habitats such as hot and acidic water springs, amid radioactive waste, deep in both the sea and the earth's crust. Some bacteria can even survive in the extreme conditions of outer space, as has already been demonstrated by scientists from the ESA (European Space Agency) and NASA.

Bacteria are so abundant that it is estimated that there are around 40 million bacterial cells in 1 g of soil and one million bacterial cells in one milliliter of fresh water. In total, there are approximately 5×10^{30} bacteria worldwide, a truly impressive number that shows that bacteria have successfully colonized our planet for billions of years.[4] However, less than half of the known species of bacteria have been cultured in the laboratory. Moreover, it is estimated that a large part of the existing bacterial species, perhaps up to 90%, have not even been scientifically described yet.

In the human body there are approximately ten times more bacterial cells than human cells, especially in the skin and digestive tract. Human cells are much larger but bacterial cells are much more abundant. Fortunately, most bacteria present in the human body are harmless or beneficial (although some pathogenic bacteria can cause infectious diseases, such as cholera, diphtheria, leprosy, syphilis, or tuberculosis).

Bacteria are very simple microorganisms and do not have a nucleus, so they are called prokaryotes (from the Greek "pro" meaning "before" and "karyon" meaning "nut" or "nucleus"). Bacteria generally have only one circular chromosome, and they do not have a separate nucleus as such. A circular chromosome has neither a beginning nor an end, which is why it has no telomeres either (from the Greek "telos" meaning "end" and "mere" meaning "part"). On the other hand, eukaryotic cells (from the Greek "eu" meaning "true" and "karyon" meaning "nut" or "nucleus") have "end parts" or telomeres because their chromosomes are not circular. The word bacteria ("stick" in Greek) was coined by German scientist Christian Ehrenberg in 1828, and French biologist Edouard Chatton created the words "prokaryote" and "eukaryote" in 1925 to distinguish organisms without a true nucleus, such as bacteria, and organisms with a nucleus, such as plants and animals.

The evolutionary success of the bacteria allowed them to colonize all parts of the planet, generating countless species of bacteria, many of which are still unknown. In fact, the evolution of these organisms, like the evolution of all other life forms, is still going on. It was first thought that bacteria had only one circular chromosome, but then bacteria were found with more chromosomes, including linear chromosomes and combinations of circular and linear chromosomes. It's really fascinating to see how life permanently experiments with multiple possibilities.[5]

Evolutionarily, prokaryotic cells (without nucleus) appeared before eukaryotic cells (with nucleus). There are other microorganisms without nucleus called archaea, less abundant and possibly of later appearance to the bacteria and that together with them form the group of prokaryotes. At the evolutionary level, it is estimated that there was a last universal common ancestor known as LUCA (*Last Universal Common Ancestor*), which must have existed about 4000 million years ago, and from which all current life forms are derived, first prokaryotes (bacteria and archaea) and then eukaryotes (including current animals and plants). All living beings have the basic genetic material with DNA from the original LUCA ancestor, with a minimum of 355 original genes made from four nucleotide bases called: adenine (A), cytosine (C), guanine (G) and thymine (T).[6]

Figure 1.1 shows the so-called phylogenetic tree of life, where the two large groups (sometimes called "domains", "kingdoms" or "empires") of prokaryotes (mainly unicellular organisms: bacteria and archaea) and eukaryotes (mainly multicellular organisms: where fungi, animals and plants appear) can be clearly observed. Biology is very complex, and evolution has taken millions

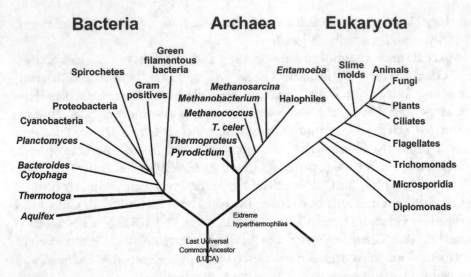

Fig. 1.1 Phylogenetic tree of life

of years to act, so it should be noted that there are also multicellular prokaryotes on the one hand, and unicellular eukaryotes on the other. However, most large eukaryotic organisms are multicellular and contain linear chromosomes with telomeres at the ends of the chromosomes, within the great phylogenetic tree of life, with the common origin of LUCA.

At the reproductive level, bacteria can be considered biologically "immortal" under ideal growing conditions. Under the best conditions, when a cell divides symmetrically it produces two daughter cells, and this process of cell division restores each cell to a young state. That is, in this type of symmetrical asexual reproduction, each offspring cell is equal to the parent cell (except for some possible mutation in cell division) but in a young state. In other words, bacteria that reproduce in this way can be considered biologically immortal. Similarly, the stem cells and gametes of multicellular organisms can also be considered "immortal", as we will see later.

Spanish microbiologists from the University of Barcelona, Ricardo Guerrero and Mercedes Berlanga, explained prokaryotic "immortality":[7]

Oddly enough, aging and death, which are the final destination of humans, were not necessary at the dawn of life, and were not for hundreds of millions of years. The classical definition of a living being as one who "is born, grows, reproduces and dies" cannot be applied in the same way to prokaryotes as to eukaryotes.

In a dividing prokaryotic cell, DNA is carried away by the membrane to which it is attached as it grows, until the cell divides to form two cells identical to the progenitor. Whenever the environment allows, prokaryotes can grow and divide without aging. Although there are variations from the general pattern, the typical cell division of bacteria occurs by "binary fission" and results in two equivalent cells.

However, not all bacteria divide symmetrically with a growth called "intercalary" that produces equal offspring cells that reproduce without aging. Guerrero and Berlanga also clarify:

> With intercalary growth, the cells, theoretically, do not die. Obviously, like all life forms, bacteria can "die" from hunger (absence of nutrients), heat (high temperature), high salt concentration, desiccation or dehydration, etc.

It should be noted that not all bacteria divide in this way. Bacteria that are asymmetrically divided by "polar" growth generate bacteria of differentiated offspring that eventually age and die.

Although we do not know many details about the origin and evolution of life, from a certain point of view we can say that: life arose to live, life did not arise to die. At least among bacteria that reproduce symmetrically and that do not age, under ideal conditions, but not among bacteria that reproduce asymmetrically, that do age.

It is obvious that death has always been present, but the first forms of life evolved in order to live, perhaps indefinitely young, under ideal conditions. However, the harsh realities of life, such as lack of food or disease, led to death, both for aging and non-aging organisms.

From Unicellular Prokaryotes to Multicellular Eukaryotes

Scientists estimate that the first organisms with a true nucleus, i.e., eukaryotes, appeared about 2 billion years ago, also descendants of the common ancestor LUCA, with the same type of DNA as all subsequent life forms on Earth. The first eukaryotic organisms were also unicellular, among them fungi, and specifically the first yeasts, which are also considered biologically "immortal".

In a research study published in the scientific journal *Cell* in 2013, a group of researchers from the United States and the United Kingdom reported the following results according to their experiments with the reproduction of the so-called fission yeast:[8]

Many unicellular organisms age: as time passes, they divide more slowly and ultimately die. In budding yeast, asymmetric segregation of cellular damage results in aging mother cells and rejuvenated daughters. We hypothesize that the organisms in which this asymmetry is lacking, or can be modulated, may not undergo aging.

Lifespan extension also occurs in mutants that have increased capacity to handle stress-related damage and in species that acquired more efficient stress resistance mechanisms. In organisms in which aging is not present, stress may trigger aging either due to an increase in the damage production rate or by changing the way damage is segregated.

The current paradigm in aging research argues that all organisms age. We have challenged this view by failing to detect aging in *S. pombe* cells grown in favorable conditions. We have shown that *S. pombe* undergoes a transition between nonaging and aging, due to asymmetric segregation of a high amount of damage. Further studies will elucidate the mechanisms underlying the transition to aging and its dependence on environmental components.

Human somatic cells show aging, dividing for a limited number of times in vitro whereas cancer cells, germ cells, and self-renewing stem cells are thought to exhibit replicative immortality… Comparative studies of aging and nonaging life strategies across single-cell species will help to clarify what determines the replicative potential and aging of cells in higher eukaryotic organisms.

The authors of the study emphasize these findings:

- *Fission yeast cells do not age under favorable growth conditions.*
- *Absence of aging is independent of the symmetry of division.*
- *Aging occurs after stress-induced asymmetric damage segregation.*
- *After stress the inheritance of aggregates correlates with aging and death.*

The unicellular yeasts were among the first eukaryotes, and it is presumed that they preserved the possibility of dividing without aging under ideal conditions. Evolution continued and around 1.5 billion years ago the first multicellular eukaryotic organisms appeared. Later, about 1.2 billion years ago sexual reproduction appears along with germ cells and somatic cells within multicellular eukaryotic organisms. (As with almost everything in biology, there are exceptions: not all multicellular eukaryotic organisms reproduce sexually.)

At the end of the nineteenth century scientists began to investigate germ cells as if they were totally different from the rest of somatic cells (from the Greek, "soma" means "body"). Basically, multicellular organisms are composed of many somatic cells, but the few germ cells are fundamental to the continuity and survival of the species. Germ cells produce gametes (eggs and sperm) for sexual reproduction. Additionally, germ cells are considered biologically immortal, which means that they do not age in the same way as somatic cells.[9] However, germ cells die when the rest of the body dies, because the body is mainly made up of somatic cells that do age.

In general, somatic cells are divided by "mitosis" (with a similar distribution of genetic material) and originate most of the body's cells.

Germ cells are divided by "meiosis" (which in sexually reproducing organisms produces eggs or sperm with half of the genetic material and then combines during fertilization between gametes).

Sexual reproduction has many advantages over single cell division, such as allowing faster evolution, but also many disadvantages, such as only requiring germ cells to be biologically immortal. From a biological point of view, somatic cells are disposable with sexual reproduction, while germ cells are not only immortal (i.e. they do not age in their own generation), but also transmit their genetic material from generation to generation through sexual reproduction.

Sexual selection of eukaryotic organisms is a type of natural selection (according to the ideas of the English naturalist Charles Darwin) in which some individuals reproduce more successfully than others in a population due to intersex selection. Sexual reproduction can be seen as an evolutionary force that does not exist in asexual populations. On the other hand, prokaryotic organisms, whose cells may have additional material or transformed due to mutations over time, reproduce through symmetrical or asymmetrical asexual reproduction. (In specific cases such as horizontal gene transfer, processes called conjugation, transformation or transduction may occur, which are somewhat similar to sexual reproduction.)

Immortal or "Negligible" Senescence Organisms

The biology and evolution of life are so fascinating and so full of surprises that today we can say, as we have insisted, that life appeared in order to live, as shown by bacteria that reproduce symmetrically under ideal conditions. In addition to prokaryotic organisms such as bacteria, there are also eukaryotic organisms such as yeasts that can be biologically immortal. Aging organisms also demonstrate this characteristic in cells that are key to their development, such as germ cells and stem cells from eukaryotic organisms that do not age either, that is, they are biologically immortal. Unfortunately, somatic cells do age and, when they die, drag with them to death the germ cells and pluripotent stem cells within the body.

Thanks to the continuous advances of science, today we also know that there are multicellular eukaryotic organisms that are biologically immortal, not only their germ cells, but also their somatic cells. Hydras are an excellent example of this ability to not age and to regenerate, and perhaps the ancient Greeks knew this when they spoke of the famous great hydras in their mythology. Its name comes from the mythological creature of the same name, from which two heads would sprout if one was cut off.

Hydra is a species of *cnidarians* that live in fresh waters. They measure a few millimetres and are predators, capturing small prey with their tentacles loaded with stinging cells. They have an amazing power of regeneration, reproduce both asexually and sexually, and are hermaphrodites. All *cnidarians* can regenerate, allowing them to recover from wounds because their cells divide continuously. A pioneering article by the American biologist Daniel Martinez published in 1998 in the scientific journal *Experimental Gerontology* states that:[10]

Senescence, a deteriorative process that increases the probability of death of an organism with increasing chronological age, has been found in all metazoans where careful studies have been carried out. There has been much controversy, however, about the potential immortality of hydra, a solitary freshwater member of the phylum *Cnidaria*, one of the earliest diverging metazoan groups. Researchers have suggested that hydra is capable of escaping aging by constantly renewing the tissues of its body. But no data have been published to support this assertion. To test for the presence or absence of aging in hydra, mortality and reproductive rates for three hydra cohorts have been analyzed for a period of four years. The results provide no evidence for aging in hydra: mortality rates have remained extremely low and there are no apparent signs of decline in reproductive rates. Hydra may have indeed escaped senescence and may be potentially immortal.

Different types of jellyfish can also be considered biologically immortal. For example, the *Turritopsis dohrnii*, or *Turritopsis nutricula*, is a species of small jellyfish that uses a form of biological transdifferentiation to replenish cells after sexual reproduction. This cycle can be repeated indefinitely, making them biologically immortal. Other similar animals include the jellyfish *Laodicea undulata* and the *sciphozoa Aurelia*. A 2015 scientific study indicates that:[11]

The genus Aurelia is one of the major contributors to jellyfish blooms in coastal waters, possibly due in part to hydroclimatic and anthropogenic causes, as well as their highly adaptive reproductive traits. Despite the wide plasticity of *cnidarian* life cycles, especially those recognized in certain *Hydroza* species, the known modifications of Aurelia life history were mostly restricted to its polyp stage. In this study, we document the formation of polyps directly from the ectoderm of degenerating juvenile medusae... This is the first evidence for back-transformation of sexually mature medusae into polyps in Aurelia sp.1. The resulting reconstruction of the schematic life cycle of Aurelia reveals the underestimated potential of life cycle reversal in scyphozoan medusae, with possible implications for biological and ecological studies.

The molecular processes that take place within these jellyfish during their remarkable transformation could become key parts of new therapies with human applicability. Japanese researcher Shin Kubota, a world expert on the so-called "immortal jellyfish", has conducted thorough research on this animal and has high hopes for what could be discovered thanks to new research. Kubota expresses his vision this way in The New York Times:[12]

> *Turritopsis* application for human beings is the most wonderful dream of mankind. Once we determine how the jellyfish rejuvenates itself, we should achieve very great things. My opinion is that we will evolve and become immortal ourselves.

The worms known as Planarias can be cut into pieces and each piece will have the ability to regenerate a complete worm. Planarias reproduce both sexually and asexually. Studies suggest that planarians appear to regenerate (i.e., heal) indefinitely, with a seemingly unlimited regenerative capacity (thanks to the continued growth of their telomeres) fueled by a population of highly proliferative adult stem cells. As described in a 2012 scientific article:[13]

> Some animals may be potentially immortal or at least very long-lived. Understanding the mechanisms that have evolved to allow some animals to be immortal may shed further light on the possibilities for alleviating aging and age-related phenotypes in human cells. These animals must have the capacity to replace aged, damaged, or diseased tissues and cells and hence use a population(s) of proliferative stem cells able to do this.
>
> Planarians have been described as "immortal under the edge of the knife", and may have an indefinite capacity to renew their differentiated tissues from a pool of potentially immortal planarian adult stem cells.

Other research suggests that lobsters do not weaken or lose fertility with age, and that older lobsters may be more fertile than younger lobsters. Their longevity may be due to telomerase, an enzyme that repairs long repeating sections of DNA sequences at the ends of chromosomes, known as telomeres. Most vertebrates express telomerase during the embryonic stages, but it is generally absent during the adult stages of life. Lobsters, unlike vertebrates, express telomerase in most adult tissues, which has been suggested to be related to their longevity.[14] However, lobsters are not immortal as they grow by molting, which requires increasing amounts of energy, and the larger the shell, the more energy is required. Over time, lobster will probably die of depletion during molt. It is also known that old lobsters stop molting, which means that the remaining shell will be damaged, infected, or crumble, causing death.

U.S. biogerontologist Caleb Finch, professor emeritus at the University of Southern California, is one of the world's experts on aging issues and comparisons between different species. Finch coined the term *"negligible senescence"* to describe species in which:[15]

> there is no evidence of physiological dysfunctions at advanced ages, no acceleration of mortality during adult life, and no recognized characteristic limit to life span.

Negligible senescence does not mean complete immortality, for there are always causes of death, such as depredation and accidents, or energetic and physical limitations, such as in the case of molting or destruction of the shell in lobsters, for example. As we have seen before, bacteria are very fragile organisms but they can live indefinitely in ideal conditions, either individually or in a colony.

There are clonal colonies or groups of genetically identical individuals, such as plants, fungi or bacteria, that have grown in a given place, all of them originating from a single ancestor by vegetative, non-sexual reproduction. Some of these clonal colonies have been alive for thousands of years. The largest known to date is a giant aquatic plant discovered in 2006 between the islands of Formentera and Ibiza:[16]

> At 100,000 years old, the Posidonia sea grass meadow was first taking root at the same time some of our earliest ancestors were creating the first known "art studio" in South Africa. It lives in the UNESCO-protected waterway between the islands of Ibiza and Formentera.

Another candidate for the world's longest-lived clonal organism is the one known as Pando, or the Trembling Giant, which arose from a single male trembling poplar (*Populus tremuloides*) located in Utah, United States. According to genetic markers it has been determined that the entire colony is part of a single living organism with a massive system of roots underground. Pando's root system is considered one of the oldest living organisms in the world, approximately 80,000 years old, and it is estimated that the plant collectively weighs more than 6600 tons, making it the heaviest living organism.[17]

Other clonal organisms more than 10,000 years old, formed by different colonies of plants and fungi that grow and reproduce asexually, have also been identified. As individual organisms, perhaps the longest lived are the "endoliths" (archaeas, bacteria, fungus, lichens, algae or amoebas) that live inside a rock, coral, exoskeleton, or in the pores between the rock's mineral grains. Many are extremophiles because they live in places that were once thought to be inhospitable to any kind of life. Endoliths (from the Greek: "inside the stone") are particularly studied by astrobiologists, who develop theories about endolytic environments on Mars and other planets as potential refuges for extraterrestrial microbial communities. In 2013, an international group of scientists reported a major scientific discovery with marine endoliths:[18]

> They report having found bacteria, fungi and viruses living a mile and a half beneath the ocean floor—such specimens, they report, appear to be millions of years old and reproduce only every 10,000 years.

There are several types of long-lived land and aquatic animals, including certain corals and sponges. In the case of long-lived trees, the most accurate estimates include the famous *Prometheus*, which in 1964 was cut down to verify its age of about 5000 years, and currently its relative *Methuselah*, which is estimated to be 4850 years old. In addition, there is another unnamed tree whose location has not been disclosed to prevent harm (it is estimated to be about 5062 years old, according to public information available in 2010).[19] All these trees are pines of the species *Pinus longaeva* and are the longest-lived individual organisms we know today. To put it in perspective, let's think that these trees were born long before the construction of the pyramids in Egypt, for example.[20]

In Wales there is a tree called the Llangernyw Yew with an estimated age between 4000 and 5000 years. It is a plant of the species *Taxus baccata*, located in the garden of a church in the town of Llangernyw, Conway.[21] In other parts of the world, from Chile to Japan there are other species of trees such as conifers and olive trees with estimated ages of 2, 3 and up to 4000 years.

A sacred fig tree, *Ficus religiosa* species, called *Jaya Sri Maha Bodhi* in Anuradhapura, Sri Lanka, is over 2300 years old, having been planted in 288 B.C. Therefore, it is the oldest human-planted tree known to date in the world, and is a direct descendant of the original *Bodhi* tree in India, under which Siddhartha Gautama, known as Buddha, sat to meditate and attained "spiritual enlightenment".[22]

Portuguese microbiologist João Pedro de Magalhães, professor at the University of Birmingham, maintains an Animal Aging and Longevity Database. This is an interesting list of organisms with a negligible senescence rate (along with estimated longevity in the wild) that includes the maximum ages known for these species so far:[23]

- Icelandic clam (*Arctica islandica*)—507 years old
- Rougheye rockfish (*Sebastes aleutianus*)—205 years old
- Red Sea Urchin (*Strongylocentrotus franciscanus*)—200 years
- Eastern Box Turtle (*Terrapene carolina*)—138 years old
- Olm (*Proteus anguinus*)—102 years old
- Blanding's turtle (*Emydoidea blandingii*)—77 years old
- Painted turtle (*Chrysemys picta*)—61 years old

In this list we could also include hydras, jellyfish, planarians, bacteria and yeasts, under ideal conditions, described above. In addition, it has recently been identified that the Greenland shark, of the species *Somniosus microcephalus*, could live 400 years according to what we know of its longevity. These are all species with negligible senescence, from which we will continue to learn a great deal in the coming years.[24]

The situation is no different in humans, as we have germ cells and pluripotent stem cells that do not age, although the rest of the body is made up of somatic cells that do age. The record for human longevity is that of Jeanne Louise Calment, who was born on 21 February 1875 and died on 4 August 1997. Calment was a French super-centenarian (centenarians are people who live more than 100 years and super centenarians are those who live more than 110 years) confirmed as the longest recorded person in history after reaching the age of 122 years and 164 days. She lived all her life in the city of Arles, in the south of France, met Vincent van Gogh, and is also the only person in history to have apparently reached the ages of 120, 121 and 122 years. Calment lived a very active life for her age, practiced fencing until the age of 85 and continued to ride a bicycle until the age of 100.[25]

There are groups of scientists studying centenarians and super-centenarians to understand more about human aging, from genetic factors to environmental factors, including nutrition.[26] However, even super-centenarians still age and suffer from senescence, so it is essential to learn from negligible senescence organisms.

Henrietta Lacks' 'Immortal' Cells

Henrietta Lacks was a tobacco farmer who was born in Virginia on August 1, 1920 and died in Maryland on October 4, 1951. Henrietta, born with the name Loretta Pleasant, came from a poor African American family and married her cousin David Lacks in Halifax, Virginia, before moving near Baltimore, Maryland, where she died of cancer.

Henrietta Lacks' story is narrated by science journalist Rebecca Skloot in her bestseller *The Immortal Life of Henrietta Lacks*, which was originally published in 2010 and remained on the best-selling list for 2 years:[27]

> Henrietta Lacks was an African American mother of five children and 31 years old when she died of cervical cancer in 1951. Without her knowledge, doctors treating her at Johns Hopkins Hospital took tissue samples from her cervix for research. They generated the first viable, miraculously productive, immortalized cell line, known as HeLa. These cells have helped in medical discoveries such as the polio vaccine or AIDS treatments.

On February 1, 1951, Lacks was treated at Johns Hopkins Hospital for a painful lump in the cervix and a bleeding from the vagina. That day she was diagnosed with cervical cancer with a tumor that appeared different from those previously seen by the examining gynecologist. Before starting treatment against the tumor, cells from the carcinoma were removed for research purposes without Henrietta's knowledge or consent (which was normal at the time). On his second visit, 8 days later, Dr. George Otto Gey took another sample of the tumor and kept part of it. It is in this second sample where the so-called HeLa cells (from the name of patient Henrietta Lacks) have their origin.

Lacks was treated with radiation therapy for several days, a common treatment for these cancers in 1951. Lacks returned for further X-ray treatment, but her condition became worse and Lacks returned on August 8 to John Hopkins Hospital, where she remained until her death. Although she received treatment and blood transfusions, she died on October 4, 1951 from kidney failure. A subsequent partial autopsy showed that the cancer had metastasized to other parts of the body.

Henrietta's tumor cells were carefully studied by Dr. Gey, who discovered that HeLa cells did something he had never seen before: they were kept alive and grew in cell culture. These were the first human cells that could be developed in a laboratory and that were biologically "immortal" (they did not die after some cell divisions), and could be used for many experiments. This represented an enormous advance for medical and biological research.

HeLa cells were used by physician and virologist Jonas Salk to develop a polio vaccine. To test Salk's new vaccine, the cells were put into rapid and massive reproduction in what is considered the first "industrial" production of human cells. Since they were put into mass production, HeLa cells have been sent to scientists around the world to conduct research on cancer, AIDS, the effects of radiation and toxic substances, genetic mapping, and countless other scientific purposes. HeLa cells have also been used to investigate human sensitivity to adhesive tape, adhesives, cosmetics and many other products that we now routinely use.

Since the 1950s, scientists have produced more than 20 tons of HeLa cells, which were also the first human cells cloned in 1955. There are more than 11,000 patents involving HeLa cells, and more than 70,000 scientific experiments have been conducted on them worldwide. Thanks to HeLa cells, gene therapies and drugs have been created to treat diseases such as Parkinson's, leukaemia, breast cancer and other cancers.[28]

HeLa cells are today the oldest human cell lineage cultured *in vitro* and are the most frequently used cells. Unlike non-cancerous cells, HeLa cells can be cultured in the laboratory constantly, which is why they are referred to as "immortal cells". Thanks to HeLa cells, we now know that other types of cancer are also biologically immortal, i.e. cancer cells do not age.

The HeLa cell line has been very successful for use in cancer research. These cells proliferate abnormally fast, even when compared to other cancer cells. During cell division, HeLa cells have an active version of telomerase, the enzyme that prevents the gradual shortening of telomeres involved in cell aging and death. Thus, as we will see in the next chapter, HeLa cells elude the so-called Hayflick limit, which is the limited number of cell divisions that most normal cells can perform before dying in cell culture.

The great tragedy of cancer, unlike other diseases, is that cancer cells do not age and also reproduce continuously. That is why cancer must be killed, and the sooner the better, because cancer does not die on its own. On the contrary, cancer continues to grow, reproduce and spread throughout the body. It could be said that the "body" becomes the food of the cancer until a "metastasis" occurs and then the entire organism dies.

Is Biological Immortality Possible?

We have seen that there are already different organisms that basically do not age, i.e. organisms that have negligible senescence. What could also be considered the "best" cells in our bodies (germ cells) do not age. We have also indicated that the "worst" cells in our bodies (cancer cells) do not age either. Therefore, the question should not be whether biological immortality is possible, because it already is.

The question, as we have already discussed, should rather be when it will be possible to stop aging in the humans.

American biologist Michael Rose, from the University of California, Irvine, an expert in theories of aging, explains how "biological immortality" is possible in his article *The Scientific Conquest of Death:*[29]

Is aging universal? Clearly not. If everything aged, the continued survival of the cells responsible for producing our sperm and eggs (the 'germline') over millions of years would have been impossible. Most of the bananas that you have eaten in your lifetime come from immortal clones produced on plantations. Even in organisms like mammals, which have germ-lines that separate very early from the rest of the body, the survival and replication of the cells responsible for producing gametes (germ cells) have proceeded for hundreds of millions of years. Life can continue indefinitely.

But even if life can propagate itself indefinitely, are there any organisms that are free of aging, living with biological immortality? I must be clear about one point concerning death: It is not true that aging is required to kill organisms kept in the laboratory. Showing that a species dies in the lab is not the same thing as showing that immortality does not occur in that species. Mechanical accidents in the lab will kill many softbodied plants, animals, and microscopic creatures. Deadly mutations can kill at any age or time. It is also impossible to keep living things free of all diseases indefinitely. Being free of aging does not imply the complete absence of death. Biological 'immortals' will often die, just not because of a systematic, endogenous, ineluctable process of self-destruction. Death is not aging. Biological immortality is not freedom from death.

Instead, the demonstration of immortality requires the finding that rates of survival and reproduction do not show aging. There are many cases where such patterns are inferred anecdotally among plants and simple animals, like sea anemones. But the best quantitative data known to me were gathered by Martinez, who studied mortality rates in Hydra, the aquatic animal that used to be a staple of high school biology. Martinez found that his Hydra showed no substantial fall in survival rates over very long periods. They still died, but not in a pattern that suggested aging. Other scientists have gathered comparable data with small animals. Some species were immortal and some were not. The immortal species reproduced without sex.

Also, invoking laws of thermodynamics as limits to life is clearly incorrect, given the evolutionary immortality of life forms. Such an invocation was always rankly amateur in any case, since these laws only apply to closed systems. Life on Earth is not a closed system. The earth receives an abundant input of energy from the sun.

Thus, some of the deepest prejudices of professional biologists concerning immortality are certainly false. Aging is not universal. There are organisms that are biologically immortal.

Rose is a pioneer in longevity research with *Drosophila melanogaster* fruit flies, to which he has managed to extend lifespan four times. In 1991, Rose published his book *Evolutionary Biology of Aging*, where he hypothesizes that aging is caused by genes that have two effects, one that occurs early in life and the other much later. Genes are favored by natural selection as a result of their benefits in youth, and costs appear much later as side effects that we identify as aging. Rose also argues that aging can be stopped at a later stage of life, as he has demonstrated through his experiments extending the life of the model organism *Drosophila melanogaster* fourfold.

Like Rose, we think aging can be slowed, stopped, and surely reversed. The "proof of concept" already exists with other organisms and now it is a question of discovering how to achieve it with humans as well. It is time to move from theory to practice.

Notes

1. https://www.amazon.com/Carl-Sagan-Cosmos-Utimate-Blu-ray/dp/B06X1F546N/
2. http://www.astromia.com/biografias/joanoro.htm
3. https://www.amazon.com/Molecular-Biology-Cell-Bruce-Alberts/dp/0815345240/
4. http://www.pnas.org/content/95/12/6578.full
5. https://microbewiki.kenyon.edu/index.php/Chromosomes_in_Bacteria:_Are_they_all_single_and_circular%3F
6. https://www.nytimes.com/2016/07/26/science/last-universal-ancestor.html
7. https://www.researchgate.net/publication/5772530_The_hidden_side_of_the_prokaryotic_cell_Rediscovering_the_microbial_world
8. http://www.cell.com/current-biology/fulltext/S0960-9822(13)00973-1
9. https://wi.mit.edu/news/immortality-germ-cells

10. http://www.sciencedirect.com/science/article/pii/S0531556597001137
11. https://www.ncbi.nlm.nih.gov/pubmed/26690755
12. http://www.nytimes.com/2012/12/02/magazine/can-a-jellyfish-unlock-the-secret-of-immortality.html
13. https://www.ncbi.nlm.nih.gov/pmc/articles/PMC3306686/
14. http://onlinelibrary.wiley.com/doi/10.1016/S0014-5793(98)01357-X/abstract
15. https://www.ncbi.nlm.nih.gov/books/NBK100401/
16. http://science.time.com/2014/02/25/worlds-oldest-things/photo/08_sussman_seagrass_0910_0753_1068px/
17. https://www.nps.gov/brca/learn/nature/quakingaspen.htm
18. https://phys.org/news/2013-08-soil-beneath-ocean-harbor-bacteria.html
19. http://www.rmtrr.org/oldlist.htm
20. https://brightly.eco/blog/oldest-tree-in-the-world
21. https://welshgiftshop.com/blogs/welsh-gift-shop/13921057-llangernyw-yew-the-oldest-tree-in-wales-the-angelystor-spirit
22. http://www.srimahabodhi.org/mahavamsa.htm
23. http://genomics.senescence.info/species/nonaging.php
24. http://www.sciencemag.org/news/2016/08/greenland-shark-may-live-400-years-smashing-longevity-record
25. https://en.wikipedia.org/wiki/Jeanne_Calment
26. https://www.amazon.com/Age-Later-Healthiest-Sharpest-Centenarians/dp/1250230853/
27. https://www.amazon.com/Immortal-Life-Henrietta-Lacks/dp/1400052181/
28. https://web.archive.org/web/20101121105322/http://hamptonroads.com/2010/05/cancer-cells-killed-her-then-they-made-her-immortal
29. http://www.imminst.org/SCOD.pdf

2

What Is Aging?

The reasons for some animals being long-lived and others short-lived, and, in a word, the causes of the length and brevity of life, call for investigation.
Aristotle, 350 BC

Aging is a disease that should be treated like any other.
Elie Metchnikoff, 1903

Aging is not natural.
María Blasco, 2016

Aging is something plastic that we can manipulate.
Juan Carlos Izpisúa Belmonte, 2016

Aging is a disease, the most common disease, and it should be aggressively treated.
David Sinclair, 2019

The scientific study of aging is relatively recent, and even more recent is the scientific study of rejuvenation. To exaggerate a bit, we could say that the modern science of aging is only a few decades old, and that the modern science of rejuvenation is only a few years old. Both endeavors have barely begun at the level of laboratory tests, first with model organisms, in order to be able to apply it to humans in due course. Fortunately, more and more people within and outside the scientific community realize that we will soon have scientific therapies available to slow aging, reverse aging, and begin rejuvenation in humans.

In the fourth century BC, the Greek philosopher Aristotle was one of the first to raise the scientific study of aging in both plants and animals. In the second century A.D., the Greek physician Galen proposed that aging began with the change and deterioration of the body from the earliest age. In the

© The Author(s), under exclusive license to Springer Nature Switzerland AG 2023
J. Cordeiro, D. Wood, *The Death of Death*, Copernicus Books,
https://doi.org/10.1007/978-3-031-28927-9_2

thirteenth century, the English philosopher and monk Roger Bacon put forward the *"wear and tear"* theory. In the nineteenth century, the ideas of the English naturalist Charles Darwin opened the door to evolutionary theories of aging, as well as to sustained discussions about programmed aging versus unprogrammed aging.[1]

Forms of Aging, More Aging and Non-aging

As we saw in the first chapter of this book, there are organisms that do not age, as well as cells that do not age either, even within the human body itself. Other organisms can regenerate many parts of their bodies, including the brain.[2] In other words, aging cannot be considered as a single or unitary process, as there are some forms of life that do not age and others that show negligible senescence.

Today we also know that there are organisms of the same species that may or may not age, depending on the type of reproduction. Generally speaking, asexual reproduction is prone to non-aging, while sexual reproduction is prone to aging, even in hermaphrodite individuals of the same species.

In addition, there are differences in the rate of aging between individuals of the same species, between female, male or hermaphrodite organisms. Females of some species have a different life expectancy than males, and the same is seen in species with hermaphrodite organisms. There are also considerable differences between the aging of members of sociable insect colonies, such as the great difference between the life expectancies of queen bees, worker bees, and drones.[3]

Environmental conditions also greatly influence life expectancy, mainly in species such as insects and invertebrates that do not control their body temperature. For example, temperature levels and the amount of food produce large changes in the life expectancy of worms and flies. Decreasing temperatures and caloric restriction increase the life expectancy of several species.

Several genes have been found that control part of the aging process, such as the discovery of genes called age-1 and daf-2 in *C. elegans* worms and FOXO genes in *Drosophila melanogaster* flies. These genes, and others discovered later, have some mammalian equivalents, so it is helpful to understand how they work with a view to controlling human aging (since today we also understand that aging can be genetically modified).

Everyone knows that there are organisms that live a short time, or a long time, although time is a relative concept. At one extreme we have some primitive insects, such as the so-called *ephemeral* ones, that only live one day or less in their adult form, and at the other extreme we have humans that can live a century or more (as well as species with negligible senescence). Today we also know that there are life forms with individuals who have survived centuries and millennia of which the potential limit of their longevity is not known.

Plants and animals also age differently, as Aristotle observed centuries ago. Animal cells and plant cells present great differences that have consequences on the aging model, or even non-aging or negligible senescence for some species, such as the so-called "perennial plants" (e.g. sequoias). Bacteria and fungi, for example, may not age, depending on their mode of reproduction, symmetric division rate, cell type and chromosomes.

There are also cells that live a short time, and others that live a long time, even within the same organism. For example, in humans, sperm have a life expectancy of 3 days (although the germ cells that produce them do not age), colon cells usually live 4 days, skin cells 2 or 3 weeks, red blood cells 4 months, white blood cells more than 1 year, and neocortex neurons usually last a lifetime. Today we also know that neurons in some parts of the brain can regenerate, unlike what was thought until recently, as there are also stem cells in different regions of the brain.[4]

Cells with circular chromosomes, as in most bacteria, are usually biologically immortal under ideal conditions; whereas cells with linear chromosomes, as in most somatic cells of multicellular organisms, are usually mortal unless they develop cancer and stop aging.

Today we know that cancer cells can become biologically immortal as a result of mutations in normal somatic cells that do age. Cancer stem cells are currently being studied to find clues about biological immortality in normal somatic cells as well. In other words, despite their malignancy, cancer cells can also help unravel the mystery of aging.

Some cancer cells produce the enzyme telomerase to increase the length of their telomeres at the end of chromosomes.[5] The somatic cells of many species do not produce telomerase in adult individuals, although in some cases they do and this allows continuous regeneration at the cellular level, as in the case of planar worms and some amphibians.

The above examples show that biology has had millions of years to experiment with different life forms, different species of organisms, different ways of reproduction, different types of sex, different forms of cells, different growth patterns, and different models of aging, including non-aging in some cases.

Romanian geriatrician Anca Iovita published her book *The Aging Gap Between Species* in 2015. Iovita begins by "finding the forest among the trees," she says:[6]

Aging is a puzzle to solve.

This process is traditionally studied in a couple of biological models like fruit flies, worms and mice. What all these species have in common is their fast aging. This is excellent for lab budgets. It is a great short-term strategy. Who has time to study species that live for decades?

But lifespan differences among species are magnitudes of order larger than any lifespan variation achieved in the lab. This is the reason for which I studied countless information resources in an attempt to gather highly specialized research into one easy-to-follow book. I wanted to see the forest among the trees. I wanted to expose the aging gap between species in an easy-to-follow and logical sequence. This book is my attempt at doing just that.

Aging is inevitable, or so I've been told. I was never the one to accept things at face value just because some authority said it. So, I began to question whether aging is the same in all species. While looking for answers, I was surprised to find out there is a lack of biological model diversity in gerontology. I was undeterred and I searched for the most obscure scientific paper on how other species age and what could set them apart.

If you ever had a pet, you already noticed that lifespans differ widely. You may have looked the same for a decade, while your dog or cat already suffered from age-related diseases. There is huge lifespan variability, both in terms of individuals belonging to the same species and among species themselves. What are the mechanisms underlying the aging gap between species?

In her book, Iovita makes an excellent review of current scientific knowledge about aging, including the huge differences between different species (from bacteria to whales), different theories of senescence, neoteny (i.e., the maintenance of youthful abilities such as regeneration in adults, from the Greek: "extended youth") and progeria (premature aging, from Greek: "to old"), and other fundamental subjects such as stem cells, cancer, telomerase and telomeres. Iovita concludes:

Aging is a plastic phenomenon. But lifespan differences among species are magnitudes of order larger than any lifespan variation achieved in the lab. This is the reason for which I studied countless information resources in an attempt to gather highly specialized research into one easy-to-follow book. I intentionally chose to write the answer to this question in plain English. Aging research is too important to hide it behind the closed doors of formal scientific jargon.

Gerontology as a science can progress by studying not only short-lived species like mice and worms, but gradually and specially negligibly senescent ones like sponges, naked mole rats, sea urchins, olms and many millennial trees. If aging is an increase in mortality rates and decrease in fertility rates, then the existence of negligible senescence species indirectly shows that aging is an accident of nature.

Long-lived species often continue to express telomerase in their adult somatic tissues allowing them to regenerate at least part of their organs. Despite their adult expression of telomerase, such species do not have a higher cancer rate. They probably developed alternative mechanisms to keep cancer at bay while increasing the control of their cells. The naked mole rat is considered a cancer-proof species, despite the abundant expression of telomerase in its somatic stem cells.

The magnitude of the project makes this book a work in progress. There are still countless species to be discovered. There are still aging experiments to be done and theories to be created. Aging is an accident of nature. And gerontology – the science of aging – was born to solve the puzzle of aging.

Another expert researcher who has made a careful study of the differences in aging between different species is Professor Steven Austad, who holds the Protective Life Endowed Chair in Healthy Aging Research at the University of Alabama at Birmingham. In 2022, Austad released his own book *Methuselah's Zoo: What Nature Can Teach Us about Living Longer, Healthier Lives*, with the following description from its publishers:[7]

Is there anything humans can learn from the exceptional longevity of some animals in the wild? In *Methusaleh's Zoo*, Steven Austad tells the stories of some extraordinary animals, considering why, for example, animal species that fly live longer than earthbound species and why animals found in the ocean live longest of all.

Austad argues that the best way we will learn from these long-lived animals is by studying them in the wild. Accordingly, he proceeds habitat by habitat, examining animals that spend most of their lives in the air, comparing insects, birds, and bats; animals that live on, and under, the ground—from mole rats to elephants; and animals that live in the sea, including quahogs, carp, and dolphins.

Humans have dramatically increased their lifespan with only a limited increase in healthspan; we're more and more prone to diseases as we grow older. By contrast, these species have successfully avoided both environmental hazards and the depredations of aging. Can we be more like them?

The Origins of the Scientific Study of Aging

At the end of the nineteenth century, when the revolutionary ideas on evolution that Darwin had proposed were still struggling to prevail in the scientific world, the German biologist August Weismann developed his theory on inheritance based on the immortality of germinal plasma in 1892. According to this theory, germplasm is the substance around which new cells develop.

This substance, constituted by the union of the sperm and the ovum, establishes a fundamental continuity that is not interrupted through the generations.[8]

This theory was also known at the time as Weismannism and established that hereditary information is only transmitted from the germ cells of the gonads (ovaries and testicles) and never from somatic cells. The idea that information, contrary to the theory of the French biologist Jean-Baptiste Lamarck, cannot pass from somatic cells to germ cells, is called Weismann's barrier.

Weismann's new theory, which anticipated the development of modern genetics, distinguished the immortality of germinal plasma from the mortality of the "soma" (body). Weismann further postulated that death is not inherent in life, but is rather a subsequent biological acquisition necessary for evolutionary development (to dispose of unfit and inferior organisms):[9]

> Death is to be looked on as an occurrence advantageous to the species as a concession to the outward conditions of life, not as an absolute necessity, essentially inherent in life itself. Death that is the end of life is by no means as is usually assumed an attribute of all organisms.
>
> Death itself, and the length of life, longer or shorter, depend entirely on adaptation. Death is not an essential attribute of living matter; it is not necessarily associated with reproduction, nor is it a necessary consequence of it.

Shortly afterwards, the Russian-French biologist Elie Metchnikoff, winner of the Nobel Prize in Physiology or Medicine in 1908, defended some similar ideas about evolution and immortality. He explained that not only were germ cells immortal, but multicellular organisms could also become immortal. At that time, it was generally considered that only unicellular organisms were probably immortal, and that multicellular organisms were not. It was then that Weismann explained that germ cells were biologically immortal, whereas somatic cells were mortal, and that death could play a role in evolution even if it was not necessary.

Metchnikoff worked with French biologist Louis Pasteur and coined the word "gerontology" (from the Greek: "study of old age"), so he is usually known as the "father" of gerontology. Metchnikoff agreed with Weismann that death is not a necessary prerequisite for life, since unicellular organisms and germ cells are potentially immortal. However, Metchnikoff did not believe that natural death could be an evolutionary advantage. According to him, "normal aging" and "natural death" almost never occur in nature. Weakened organisms are eliminated by external causes (predation, disease, accidents, competition) with a minimal chance of "naturally aging" or dying naturally. If aging and natural death almost never occur in nature, then natural selection cannot operate on them, much less select them to generate a competitive advantage.[10]

A few years later, French-American biologist Alexis Carrel, winner of the Nobel Prize in Physiology or Medicine in 1912, conducted experiments that suggested that somatic cells could also live indefinitely. Carrel continued researching longevity, immortal cells, tissue cultures, and organ transplants until his death in 1944. Some time later, the American micro biologist Leonard Hayflick discovered in 1961 that the somatic cells of multicellular organisms only divided a certain number of times before dying. Hayflick confirmed that germ cells (and cancerous cells, even working with HeLa cells) were biologically immortal, but that somatic cells were mortal and died after a certain number of divisions, a number that depended on the type of cell and organism, but in no case reached 100 divisions per cell. That discovery is known today as the Hayflick Limit.[11] (It is now believed that Carrel's experiments were flawed. You can learn more about the trailblazing work of Carrel and his long-time collaborator, the celebrated aviation pioneer Charles Lindbergh, in the book by David Friedman, *The Immortalists: Charles Lindbergh, Dr. Alexis Carrel, and Their Daring Quest to Live Forever.*[12]).

The history of scientific advances in aging research during the twentieth century is truly exciting. We went from mainly conceptual theories to real experiments, some of which were erroneous and unreproducible. Scientists from Germany, Russia, France, and the United States were among the top leaders in aging research in the last century. Moldovan-Israeli researcher Ilia Stambler carefully detailed all these stories in his book *A History of Life-Extensionism in the Twentieth Century*. Stambler describes at the beginning the four major chapters of his extensive book, published in 2014:[13]

This work explores the history of life-extensionism in the 20th century. The term life-extensionism is meant to describe an ideological system professing that radical life extension (far beyond the present life expectancy) is desirable on ethical grounds and is possible to achieve through conscious scientific efforts. This work examines major lines of life-extensionist thought, in chronological order, over the course of the 20th century, while focusing on central seminal works representative of each trend and period, by such authors as Elie Metchnikoff, Bernard Shaw, Alexis Carrel, Alexander Bogomolets and others. Their works are considered in their social and intellectual context, as parts of a larger contemporary social and ideological discourse, associated with major political upheavals and social and economic patterns. The following national contexts are considered: France (Chapter 1), Germany, Austria, Romania and Switzerland (Chapter 2), Russia (Chapter 3), the US and UK (Chapter 4).

This work pursues three major aims. The first is to attempt to identify and trace throughout the century several generic biomedical methods whose development or applications were associated with radical hopes for life-extension. Beyond mere hopefulness, this work argues, the desire to radically prolong human life often constituted a formidable, though hardly ever acknowledged, motivation for biomedical research and discovery. It will be shown that novel fields of biomedical science often had their origin in far-reaching pursuits of radical life extension. The dynamic dichotomy between reductionist and holistic methods will be emphasized.

The second goal is to investigate the ideological and socio-economic backgrounds of the proponents of radical life extension, in order to determine how ideology and economic conditions motivated the life-extensionists and how it affected the science they pursued. For that purpose, the biographies and key writings of several prominent longevity advocates are studied. Their specific ideological premises (attitudes toward religion and progress, pessimism or optimism regarding human perfectibility, and ethical imperatives) as well as their socioeconomic conditions (the ability to conduct and disseminate research in a specific social or economic milieu) are examined in an attempt to find out what conditions have encouraged or discouraged life-extensionist thought.

The third, more general, aim is to collect a broad register of life-extensionist works, and, based on that register, to establish common traits and goals definitive of life-extensionism, such as valuation of life and constancy, despite all the diversity of methods and ideologies professed. This work will contribute to the understanding of extreme expectations associated with biomedical progress that have been scarcely investigated by biomedical history.

Theories of Aging in the Twenty-First Century

Despite the great advances of the twentieth century, there is still no universally accepted theory of aging. In fact, a large number of theories are currently competing, which can be divided in many ways. For example, in a course at the University of California, Berkeley, four major groups were considered: molecular theories, cellular theories, systemic theories, and evolutionary theories, each group in turn with three or more theories within the group. In total, more than a dozen theories can be classified into these four major groups: codon-restriction, error catastrophe, somatic mutation, dedifferentiation, gene regulation, wear and tear, free radical, apoptosis, senescence, rate of living, neuroendocrine, immunological, disposable soma, antagonistic pleiotropy, and mutation accumulation.[14]

The aforementioned Portuguese microbiologist João Pedro de Magalhães studies damage-based aging theories and programmed aging theories, which is also a standard classification.[15] Some biologists make a great division between mainly genetic and non-genetic theories. Others talk about evolutionary theories and physiological theories (divided into programmed and stochastic, or unprogrammed). The common point is that more and more scientists are realizing that we must systematically research aging, as evidenced by the following *Scientists' Open Letter on Aging Research* signed in 2005 by several respected scientists from around the world:[16]

Aging has been slowed and healthy lifespan prolonged in many disparate animal models (C. elegans, Drosophila, Ames dwarf mice, etc.). Thus, assuming there are common fundamental mechanisms, it should also be possible to slow aging in humans.

Greater knowledge about aging should bring better management of the debilitating pathologies associated with aging, such as cancer, cardiovascular disease, type II diabetes, and Alzheimer's. Therapies targeted at the fundamental mechanisms of aging will be instrumental in counteracting these age-related pathologies.

Therefore, this letter is a call to action for greater funding and research into both the underlying mechanisms of aging and methods for its postponement. Such research may yield dividends far greater than equal efforts to combat the age-related diseases themselves. As the mechanisms of aging are increasingly understood, increasingly effective interventions can be developed that will help prolong the healthy and productive lifespans of a great many people.

The discussion on aging has been increasing and has become globalised, from Russia through China to the United States. For example, a group of Russian scientists published in 2015 an article entitled "Theories of Aging: An Ever-Evolving Field" in the scientific journal *Acta Naturae*, where they explain:[17]

Senescence has been the focus of research for many centuries. Despite significant progress in extending average human life expectancy, the process of aging remains largely elusive and, unfortunately, inevitable. In this review, we attempted to summarize the current theories of aging and the approaches to understanding it.

In another part of the world, an American scientist of Chinese origin, Dr. Kunlin Jin of the University of North Texas Medical Science Center, published in 2010 an article entitled "Modern Biological Theories of Aging" in the scientific journal *Aging and Disease*, where he indicates:[18]

Despite recent advances in molecular biology and genetics, the mysteries that control human lifespan are yet to be unraveled. Many theories, which fall into two main categories: programmed and error theories, have been proposed to explain the process of aging, but neither of them appears to be fully satisfactory. These theories may interact with each other in a complex way. By understanding and testing the existing and new aging theories, it may be possible to promote successful aging.

Faced with this flood of theories, some old and some new, Aubrey de Grey began working systematically from the end of the twentieth century to compile all the information in an inclusive system on aging. De Grey first studied computer science and computing at Cambridge University, developing his vision more as an engineer or technologist than as a biologist or doctor. His approach to life extension is called SENS (*Strategies for Engineered Negligible Senescence*). It was in 2002 when he first presented those ideas in an article published along with other well-known physicians and biologists, such as Bruce Ames, Julie Andersen, Andrzej Bartke, Judith Campisi, Christopher Heward, Orger McCarter and Gregory Stock.[19]

The key meaning of the term "SENS" is the engineering of medical therapies to reverse biological aging in humans so that we can continue to accumulate years of age while remaining biologically young. To that end, de Grey did a thorough study of the available research on aging and realized that there are seven main types of damage related to the aging process. He also discovered that all of these types of damage have been known for at least 1982, that is, for several decades.

Biology has made immense progress since then, but scientists haven't discovered any new types of damage, according to de Grey. This suggests that we already know the key problems that combine to create the fragility and vulnerability to disease that we associate today with old age. The new approach is to attack the damage through bioengineering. It differs from gerontology, which seeks to alter metabolism, and from geriatrics, which seeks to alleviate pathology. Figure 2.1 shows the SENS strategy.

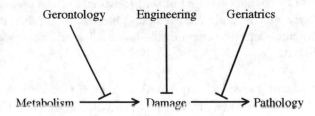

Fig. 2.1 SENS rejuvenation biotechnology research strategies. (Source: Aubrey de Grey 2008)

What are these seven causes of senescence? The seven lethal causes? They all occur at the microscopic level, inside and outside the cells. A little damage won't usually hurt you, but that damage builds up over the years at an accelerated rate, which is why people become fragile and die. In his book *Ending Aging: The Rejuvenation Breakthroughs That Could Reverse Human Aging in Our Lifetime*, de Grey explains these seven causes:[20]

1. Intracellular waste
2. Intercellular waste
3. Nucleus mutations
4. Mitochondrial mutations
5. Stem cells loss
6. Increase in senescent cells
7. Increase of intercellular protein links

Moreover, of critical importance, his book also outlines research programs to deal with each of these seven causes.

When de Grey originally put forward his ideas, many people called him names from quack to crazy. Many "experts" attacked him, asserting that his ideas had no scientific basis. The discussion reached the prestigious Technology Review of the Massachusetts Institute of Technology (MIT) in 2005, where the editor challenged, under a promise of $20,000, the first person to prove SENS "so wrong that it is unworthy of learned debate".[21] To this end, a jury was established with five prestigious scientists and doctors (Rodney Brooks, Anita Goel, Vikram Kumar, Nathan Myhrvold and Craig Venter) who would evaluate the criticisms of Aubrey de Grey's ideas. Despite all the publicity and money involved, the criticisms seemed more personal attacks than consistent arguments against SENS strategies. After several months and multiple attempts, the prize was declared void, as no one could prove that Grey's ideas were false, which did not prevent some "experts" from attacking it on the basis of personal prejudices.[22]

The world has changed a lot since 2005. There have been great scientific advances in recent years that reinforce rather than contradict Aubrey de Grey's original ideas. An article in the Smithsonian scientific journal in 2017 mentions one of the articles written against de Grey in the MIT Technology Review entitled "Life Extension Pseudoscience and the SENS Plan":[23]

The nine co-authors, all senior gerontologists, took stern issue with de Grey's position. "He's brilliant, but he had no experience in aging research," says Heidi Tissenbaum, one of the paper's signatories and a professor of molecular, cell and cancer biology at the University of Massachusetts Medical School. "We were alarmed, since he claimed to know how to prevent aging based on ideas, not on rigorous scientific experimental results."

More than a decade later, Tissenbaum now sees SENS in a more positive light. "Kudos to Aubrey," she says diplomatically. "The more people talking about aging research, the better. I give him a lot of credit for bringing attention and money to the field. When we wrote that paper, it was just him and his ideas, no research, nothing. But now they are doing a lot of basic, fundamental research, like any other lab."

Although some still call de Grey a quack and crazy, there are more and more positive results supporting his early claims. De Grey co-founded the Methuselah Foundation in 2003, which created the Methuselah Mouse Prize to encourage research to radically delay and even reverse aging. The Methuselah Mouse Prize, or simply MPrize, owes its name to Methuselah, the patriarch in the Bible who supposedly lived almost a thousand years. Thanks to this award and other incentives, the life of mice has been significantly expanded. For example, mice that live in wild conditions one year in nature, and in the laboratory between two and three years, have come to live almost five years with various treatments.

Using different types of treatments, scientists have been able to increase the life expectancy of mice by 40%, 50% and even more. Hopefully the prize will continue and soon we will be able to talk about mice that double and triple their average life expectancy.

De Grey also co-founded the SENS Research Foundation in 2009, which aims to "redefine the way the world researches and treats age-related ill health". It promotes "in situ repair of living cells and extracellular material", an approach that contrasts with the more traditional geriatric medicine of specific diseases and pathologies, and with that of biogerontology in the intervention in metabolic processes. The SENS Foundation finances research and promotes dissemination and education to speed up the various research programs in regenerative medicine. According to the SENS approach, each of the seven fundamental damages can be treated with one specific strategy, known as RepleniSENS, OncoSENS, MitoSENS, ApoptoSENS, GlycoSENS, AmyloSENS and LysoSENS. Several of these treatments are already being applied, and some have been used to boost startups seeking anti-aging and rejuvenation therapies.[24]

In his article "Undoing Aging with Molecular and Cellular Damage Repair", published in 2017 by BBVA OpenMind in the book *The Next Step: Exponential Life*, de Grey explains:[25]

SENS is a hugely radical departure from prior themes of biomedical gerontology, involving the bona fide reversal of aging rather than its mere retardation. By virtue of a painstaking process of mutual education between the fields of biogerontology and regenerative medicine, it has now risen to the status of an acknowledged viable option for the eventual medical control of aging. I believe that its credibility will continue to rise as the underlying technology of regenerative medicine progresses.

In an interview given in Madrid during the first International Longevity and Cryopreservation Summit that we organized at the Spanish National Research Council (CSIC) in Madrid, de Grey summarized the advances in the SENS strategy. The interviewers came to the following conclusions after seeing the enormous changes that had taken place over the last decade:[26]

There is much to be optimistic about. The ideas proposed by SENS over a decade ago, widely criticized in the past, are now being eagerly explored by researchers as it becomes ever more apparent that the aging processes are amenable to intervention. What was mocked just over a decade ago is now becoming an accepted approach to treating age-related diseases, as the results continue to mount up in support of a repair-based approach to aging.

However, we still require more knowledge about several age-related damages in order to progress to clinical trials in humans. This is why supporting fundamental studies on the main mechanisms of aging should remain the number one priority for our community.

With the formation in 2022 of LEV (Longevity Escape Velocity) Foundation, Aubrey de Grey envisions an acceleration the approaches pioneered by SENS:[27]

Longevity Escape Velocity (LEV) Foundation exists to proactively identify and address the most challenging obstacles on the path to the widespread availability of genuinely effective treatments to prevent and reverse human age-related disease.

LEV Foundation's first flagship research project started in January 2023, and takes to a new level some ideas first considered when establishing the Methuselah Mouse Prize:[28]

LEV Foundation's flagship research program is a sequence of large mouse lifespan studies, each involving the administration of (various subsets of) at least four interventions that have, individually, shown promise in extending mean and maximum mouse lifespan and healthspan.

We focus on interventions that have shown efficacy when begun only after the mice have reached half their typical life expectancy, and mostly on those that specifically repair some category of accumulating, eventually pathogenic, molecular or cellular damage.

The Causes and Pillars of Aging

In addition to the visionary and revolutionary work of Aubrey de Grey, other scientists are trying to systematize our current understanding of aging and how to treat it. In 2000, a couple of American oncologists, Douglas Hanahan and Robert Weinberg, wrote a provocative article in the prestigious scientific journal *Cell* that has helped to organize our knowledge about cancer. Under the title "The Causes of Cancer," the authors argue that all cancers share six common traits ("causes" or "*hallmarks*") that govern the transformation of normal cells into cancer cells (malignant or tumor cells). By 2011, the article had become the most cited in the history of *Cell*, and the authors published an update proposing four additional causes.

Building on the success of the previous article, a group of five European scientists published in 2013 an article entitled "The Hallmarks of Aging" in the same magazine *Cell*. The authors are the Spanish Carlos López-Otín (from the University of Oviedo), María Blasco and Manuel Serrano (from Spanish National Cancer Research Centre, CNIO, in Madrid), the English Linda Partridge (from the Max Planck Institute for Biology of Aging in Germany) and the Austrian Guido Kroemer (from the University of Paris V René Descartes in France), who write:[29]

Aging is characterized by a progressive loss of physiological integrity, leading to impaired function and increased vulnerability to death. This deterioration is the primary risk factor for major human pathologies, including cancer, diabetes, cardiovascular disorders, and neurodegenerative diseases. Aging research has experienced an unprecedented advance over recent years, particularly with the discovery that the rate of aging is controlled, at least to some extent, by genetic pathways and biochemical processes conserved in evolution. This Review enumerates nine tentative hallmarks that represent common denominators of aging in different organisms, with special emphasis on mammalian aging. These hallmarks are: genomic instability, telomere attrition, epigenetic alterations, loss of proteostasis, deregulated nutrient sensing, mitochondrial dysfunction, cellular senescence, stem cell exhaustion, and altered intercellular communication. A major challenge is to dissect the interconnectedness between the candidate hallmarks and their relative contributions to aging, with the final goal of identifying pharmaceutical targets to improve human health during aging, with minimal side effects.

Aging, which we broadly define as the time-dependent functional decline that affects most living organisms, has attracted curiosity and excited imagination throughout the history of humankind. However, it has only been 30 years since a new era in aging research was inaugurated following the isolation of the first long-lived strains in Caenorhabditis elegans (*C. elegans*). Nowadays, aging is subjected to scientific scrutiny based on the ever-expanding knowledge of the molecular and cellular bases of life and disease. The current situation of aging research exhibits many parallels with that of cancer research in previous decades.

Scientists classify these nine causes of aging into three major categories, as shown in Fig. 2.2. Above are so-called "primary causes" (genomic instability, telomere shortening, epigenetic alterations and loss of proteostasis), which are considered to be the underlying causes of cell damage. At the centre are the "antagonistic causes" (deregulated nutrient sensing, mitochondrial dysfunction and cell senescence), considered as part of the compensatory or antagonistic responses to damage. These responses initially mitigate the damage, but if they are chronic or exacerbated, they can become harmful. Below are the "integrative causes" (stem cell depletion and altered intercellular communication), the end result of the two previous groups, and the factors directly responsible for the functional decline associated with aging.

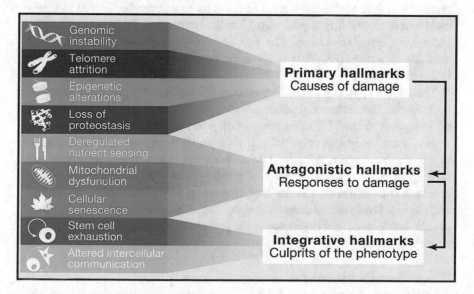

Fig. 2.2 Functional interconnections between the hallmarks of aging. (Source: Carlos López-Otín et al. 2013)

The article ends with the following conclusions and perspectives:

Defining hallmarks of aging may contribute to building a framework for future studies on the molecular mechanisms of aging and designing interventions to improve human healthspan… We surmise that ever more sophisticated approaches will eventually resolve many of the pending issues. Hopefully, these combined approaches will allow a detailed understanding of the mechanisms underlying the hallmarks of aging and will facilitate future interventions for improving human healthspan and longevity.

One year after the previous article, a group of American scientists, with the support of the U.S. National Institutes of Health, published "Aging: a common driver of chronic diseases and a target for novel interventions" in the same scientific journal Cell. The authors explain that instead of "attacking" disease by disease, it is better to directly "attack" aging itself, which is the cause of all related diseases:[30]

Mammalian aging can be delayed with genetic, dietary and pharmacologic approaches. Given that the elderly population is dramatically increasing and that aging is the greatest risk factor for a majority of chronic diseases driving both morbidity and mortality, it is critical to expand Geroscience research directed at extending human healthspan.

The goal of slowing aging has fascinated humankind for millennia, but only recently acquired credibility. Recent findings that aging can be delayed in mammals raise the possibility of prolonging human healthspan. There is near consensus among aging researchers that this is possible, but only if resources are available to accomplish goals in areas ranging from basic biology to translational medicine.

The current approach to treating chronic diseases is inadequate and fragmentary. By the time chronic diseases are diagnosed, much damage is done and undoing it is difficult. While understanding the unique features of any given disease is laudatory and potentially of therapeutic value, approaches to understand a common cause, aging, will be uniquely important. If we can understand how aging enables disease, it may be possible (and even easier) to target this common component of disease. Targeting aging may allow early intervention and damage avoidance, maintaining vigor and activity, while offsetting the economic burdens of a burgeoning aging population hampered by multiple chronic diseases.

The authors also describe what they call seven "pillars" of aging. According to Chilean-American scientist Felipe Sierra, director of the division of Aging

Biology at the National Institute on Aging in the United States, those seven pillars are:[31]

1. Inflammation
2. Adaptation to stress
3. Epigenetics and regulatory RNA
4. Metabolism
5. Macromolecular damage
6. Proteostasis
7. Stem cells and regeneration

Another author of the article, U.S. biologist Brian Kennedy, then president of the Buck Institute for Research on Aging in California, concludes that:[32]

What has come out of our work is a keen understanding that the factors driving aging are highly intertwined and that in order to extend healthspan we need an integrated approach to health and disease with the understanding that biological systems change with age.

With another perspective, the Spanish biologist Ginés Morata, expert in *Drosophila melanogaster* flies at the "Severo Ochoa" Centre for Molecular Biology in Madrid, explains that:[33]

Death is not inevitable. Bacteria do not die. Polyps don't either; they grow and generate a new one. Part of our germ cells are perpetuated in our children and so on. That is why a part of each one of us is immortal.

A type of worm, a nematode, has been made to live seven times longer by manipulating the genes involved in its aging. If we applied this technology to humans, we could live 350 or 400 years. Of course, you can't research with human material, but it's not unthinkable that one day we'll reach that longevity. In 50, 100 or 200 years the possibilities will be so great that it is hard to imagine what will happen. We may have wings and be able to fly, or measure four meters... It will be humanity that decides what its future is going to be.

American biogerontologist Michael West, an expert on stem cells and telomeres, author of several books on aging and possible rejuvenation, also agrees:[34]

Still resident in the human body are potential heirs of our immortal legacy, cells that have the potential to leave no dead ancestors; cells from a lineage called the germ-line. These cells have the ability for immortal renewal as demonstrated by the fact that babies are born young, and those babies have the potential to some-day make their own babies, and so on, forever...

In August 2022, a distinguished group of longevity researchers published an article in *Aging* that proposed adding five more hallmarks of aging to the set of nine originally listed in 2013: compromised autophagy, splicing dys-regulation, microbiome disturbance, altered mechanical properties, and inflammation.[35]

In January 2023, the same five authors who had published the original "Hallmarks of Aging" in 2013, wrote a new article to review how much we have advanced in one decade. In the same *Cell* journal, they stated:[36]

Since 2013, when the first edition of the hallmarks of aging was published in *Cell*, close to 300,000 articles dealing with this subject have been published, which is as many as during the preceding century. Hence, time has become ripe for a new edition of the hallmarks of aging incorporating the main knowledge obtained a decade on.

Indeed, it seems that aging research has been growing exponentially, with hundreds of researchers joining the field and writing thousands of articles every year. The five authors also proposed a more complete view of the current understanding of the "Hallmarks of Aging", which will surely continue to improve as more resources going into the field. In fact, as some articles have noticed: "A decade of research expands the Hallmarks of Aging from nine to twelve."[37] Figure 2.3 highlights the updated hallmarks of aging and their rela-tionships, according to these authors.

After considering so many different theories, strategies, causes and pillars of aging, what can we conclude? It is wise to separate the question of the details of various theories from the important principles that all these theories share. The definition of aging in the prestigious *Encyclopaedia Britannica* puts it well. That definition starts like this:[38]

Aging is the sequential or progressive change in an organism that leads to an increased risk of debility, disease, and death. Aging takes place in a cell, an organ, or the total organism with the passage of time. It is a process that goes on over the entire adult life span of any living thing...

Regardless of the definition we use, there is a great deal of agreement on key terms and ideas. There are also two important points to consider on which there is a growing consensus:

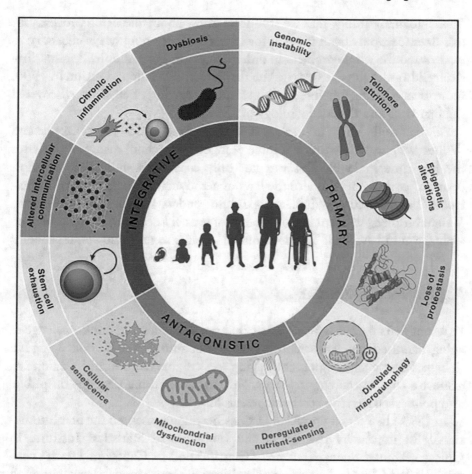

Fig. 2.3 Hallmarks of aging: an expanding universe. (https://ars.els-cdn.com/content/image/1-s2.0-S0092867422013770-gr1_lrg.jpg, Source: Carlos López-Otín et al. 2023, https://ars.els-cdn.com/content/image/1-s2.0-S0092867422013770-gr1_lrg.jpg)

- Aging occurs gradually, i.e. during a substantial part of the body's vital period. It is an essentially dynamic and sequential process, rather than just happening at a single moment, so that damage can also be attacked sequentially.
- Aging is not considered today something biologically "inevitable" or even "irreversible", rather we now know that it is a "plastic" and "flexible" process that we can manipulate. In this sense, the Handbook of the Biology of Aging also makes no reference to it being an "inevitable" process, and specifically admits the possibility that there are cells and organisms that do not age; nor is the process called "irreversible", as the Handbook speaks of the possibility that the damage can be repaired.[39]

There are many things that we still don't know about the aging process, but this doesn't stop us from moving towards a cure. It's not always necessary to understand the whole problem in order to solve it. For example, English physician Edward Jenner developed the first effective smallpox vaccine in 1796, more than a century before Dutch scientist Martinus Beijerinck discovered the first virus and founded virology in 1898.

Another well-known example is that of the American brothers Orville and Wilbur Wright, who with only three years of high school managed to fly for the first time in 1903. This was not only considered impossible by most "experts" at the time, but also the laws of aerodynamics were not well understood. The more educated scientists did not understand them, much less the Wright brothers, who could barely boast of formal knowledge. But, as Galileo Galilei would have said: "eppur si muove", that is to say, "and yet, it moves".

Aging as a Disease

In recent years there has been a great change of mentality in our knowledge of aging, and there are even scientists who are beginning to affirm that aging is a disease. Fortunately, in that case, aging is a curable disease, and we hope to achieve a cure in the coming years, although everything depends on public and political support to accelerate research.

In 1893, the French physician Jacques Bertillon presented the first international list to classify diseases at the International Statistical Institute in Chicago, United States. That first "Classification of Causes of Death" contained only 44 "causes," based on a classification system used in Paris, although it later expanded to nearly 200 when the first International Conference for the Classification of Causes of Death was held in 1900. These initial attempts at classification were first adopted by the League of Nations after World War I and then by the World Health Organization (WHO) at the end of World War II.[40]

The WHO took over the classification in 1948 with the sixth edition, the first to also include causes of morbidity. The list is now called the International Statistical Classification of Diseases and Related Health Problems, also known simply as the International Classification of Diseases (ICD). The ICD determines the classification and coding of diseases and a wide variety of signs, symptoms, social circumstances, and external causes of disease. The system is designed to promote international comparison of the collection, processing, classification and presentation of these statistics.

The most recent edition of the ICD is the eleventh, ICD-11, released in June 2018.[41] Over the preceding two decades, the ICD-10 was the internationally recognized list in force, albeit with some local modifications in certain countries. In a public suggestion period for the WHO in 2017, several activists, including the authors of this book, supported including aging as a disease, or at least initiating scientific research. Thanks to the input of the people who welcomed our proposal, the WHO agreed to include "healthy aging" in its overall work programme for 2019–2023, although it does not yet formally include aging as a disease.[42]

During the last century, some conditions that were considered diseases ceased to be diseases, just as others that were not diseases became diseases. A group of international researchers (the Belgian Sven Bulterijs, the Swede Victor C. E. Björk, and the English Raphaella S. Hull and Avi G. Roy) published in 2015 the article "It is time to classify biological aging as a disease" in the scientific journal *Frontiers in Genetics*, where they explain:[43]

What is considered to be normal and what is considered to be diseased is strongly influenced by historical context. Matters once considered to be diseases are no longer classified as such. For example, when black slaves ran away from plantations they were labeled to suffer from drapetomania and medical treatment was used to try to "cure" them. Similarly, masturbation was seen as a disease and treated with treatments such as cutting away the clitoris or cauterizing it. Finally, homosexuality was considered a disease as recently as 1974. In addition to the social and cultural influence on disease definition, new scientific and medical discoveries lead to the revision of what is a disease and what is not. For example, fever was once seen as a disease in its own right but the realization that different underlying causes would lead to the appearance of fever changed its status from disease to symptom. Conversely, several currently recognized diseases, such as osteoporosis, isolated systolic hypertension, and senile Alzheimer's disease, were in the past ascribed to normal aging. Osteoporosis was only officially recognized as a disease in 1994 by the World Health Organization.

Traditionally, aging has been viewed as a natural process and consequently not a disease. This division may have, in part, originated as a way of establishing aging as an independent discipline of research. Some authors go as far as to create a division between intrinsic aging processes (termed primary aging) and diseases of old age (termed secondary aging). For example, photoaging, the accelerated deterioration of skin as a result of UV rays during one's lifetime, is considered by dermatologists as a condition leading to pathology. In contrast, chronological skin aging is accepted as the norm. As well as being seen as separate from disease, aging is looked at as a risk factor for developing disease. Interestingly, the so-called "accelerated aging diseases" such as Hutchinson-

Gilford Progeria Syndrome, Werner syndrome, and Dyskeratosis Congenita are considered diseases. Progeria is considered a disease but yet when the same changes happen to an individual 80 years older they are considered normal and unworthy of medical attention.

These researchers mention the specific case of progeria, an extremely rare childhood genetic disease characterized by premature and accelerated aging in children between the first and second years of life. This rare condition is suffered by one in seven million live newborns. Since progeria is a genetic disease (due to mutations in a gene identified as LMNA), it is hoped that someday there will be a cure thanks to gene therapies. However, there is currently no cure or treatment for this accelerated aging disease, and progeria patients live an average of just 13 years (some patients may live to just over 20 but with a physiognomy of almost 100).

Bulterijs, Björk, Hull and Roy continue their article citing several successful research and studies on model animals, and the high costs of not yet doing so with humans (costs both at the level of the individual and society as a whole):

"In short, not only does aging lend itself to be characterized as a disease, but the advantage of doing so is that, by rejecting the seeming fatalism of the label "natural," it better legitimizes medical efforts to either eliminate it or get rid of those undesirable conditions associated with it." The goal of biomedical research is to allow people to be "as healthy as possible for as long as possible". Having aging recognized as a disease would stimulate grant-awarding bodies to increase funding for aging research and develop biomedical procedures to slow the aging process. Indeed, Engelhardt states that calling something a disease involves the commitment to medical intervention. Furthermore, having a condition recognized as a disease is important to have treatment refunded by health insurance providers.

During the last 25 years, by targeting the underlying processes of aging biomedical scientists have been able to improve the health and lifespan of model organisms, from worms and flies, to rodents and fish. We can now consistently improve the lifespan of C. elegans by more than tenfold, more than double the lifespan of flies and mice, and improve the lifespan of rats and killifish by 30 and 59%, respectively. Currently, our treatment options for the underlying processes of aging in humans are limited. However, with current progress in the development of geroprotective drugs, regenerative medicine, and precision medicine interventions, we will soon have the potential to slow down aging. Finally, we should note that recognizing aging as a disease would shift anti-aging therapies from the Federal Drug Administration's (FDA) regulations for cosmetic medicine to the more rigorous regulations for disease treatment and prevention.

We believe that aging should be seen as a disease, albeit as a disease that is a universal and multisystemic process. Our current healthcare system doesn't recognize the aging process as the underlying cause for the chronic diseases affecting the elderly. As such, the system is setup to be reactionary and therefore about 32% of total Medicare spending in the United States goes to the last 2 years of life of patients with chronic illnesses, without any significant improvement to their quality of life. Our current healthcare system is untenable both from a financial and health and well-being prospective. Even minimal attenuation of the aging process by accelerating research on aging, and development of geroprotective drugs and regenerative medicines, can greatly improve the health and well-being of older individuals, and rescue our failing healthcare system.

A few months later, other researchers wrote an article in the same scientific journal titled "Classification of Aging as a Disease in the Context of ICD-11," where they explain:[44]

Aging is a complex continuous multifactorial process leading to loss of function and crystalizing into the many age-related diseases. Here, we explore the arguments for classifying aging as a disease in the context of the upcoming World Health Organization's 11th International Statistical Classification of Diseases and Related Health Problems (ICD-11), expected to be finalized in 2018. We hypothesize that classifying aging as a disease with a "non-garbage" set of codes will result in new approaches and business models for addressing aging as a treatable condition, which will lead to both economic and healthcare benefits for all stakeholders. Actionable classification of aging as a disease may lead to more efficient allocation of resources by enabling funding bodies and other stakeholders to use quality-adjusted life years (QALYs) and healthy-years equivalent (HYE) as metrics when evaluating both research and clinical programs. We propose forming a Task Force to interface the WHO in order to develop a multidisciplinary framework for classifying aging as a disease with multiple disease codes facilitating for therapeutic interventions and preventative strategies.

The recognition of a condition or a chronic process as a disease is an important milestone for the pharmaceutical industry, academic community, healthcare and insurance companies, policy makers, and individual, as the presence of a condition in disease nomenclature and classification greatly impacts the way it is treated, researched and reimbursed. However, achieving a satisfactory definition of disease is challenging, primarily due to the vague definitions of the state of health and disease. Here, we explore the potential benefits of recognizing aging as a disease in the context of current socioeconomic challenges and recent biomedical advances.

Finally, the WHO approved ICD-11 with classification of age-related diseases in 2018, but not aging itself as a disease, which is now under consideration for ICD-12. Classifying aging as a disease will greatly contribute to curing the disease itself. In addition, it will channel the enormous resources to the causes and not to the symptoms of aging. Public and private funds must be focused on the preceding cure and not on the subsequent illness. The benefits of being healthy and young are multiplied by each individual for the whole of society. The benefits would be enormous as a whole. Treating aging as a disease will allow for increased levels of research and funding, as well as identifying a clear target for the medical, pharmaceutical and insurance industries.

It is a great opportunity, as the anti-aging and rejuvenation industry has the potential to become the world's largest industry very soon. That is why more and more scientists are starting to consider aging as a disease, like the Australian biologist David Sinclair, who wrote in his 2019 bestseller *Why We Age – and Why We Don't Have To*:[45]

I believe that aging is a disease. I believe it is treatable. I believe we can treat it within our lifetimes. And in doing so, I believe, everything we know about human health will be fundamentally changed.

Notes

1. http://www.ndhealthfacts.org/wiki/Aging
2. https://www.sciencedaily.com/releases/2009/07/090701131314.htm
3. https://www.ncbi.nlm.nih.gov/pmc/articles/PMC2527632/
4. https://www.livescience.com/33179-does-human-body-replace-cells-seven-years.html
5. https://www.ncbi.nlm.nih.gov/pmc/articles/PMC7465155/
6. https://www.amazon.com/Aging-Gap-Between-Species/dp/1517484812/
7. https://www.amazon.com/Methuselahs-Zoo-Nature-Living-Healthier/dp/0262047098/
8. http://www.esp.org/books/weismann/germ-plasm/facsimile/
9. http://www.longevityhistory.com/read-the-book-online/#_edn1119
10. http://www.longevityhistory.com/the-legacy-of-elie-metchnikoff/
11. https://www.ncbi.nlm.nih.gov/pubmed/13905658
12. https://www.harpercollins.com/products/the-immortalists-david-m-friedman

13. https://www.amazon.com/History-Life-Extensionism-Twentieth-Century/dp/1500818577
14. http://mcb.berkeley.edu/courses/mcb135k/BrianOutline.html
15. http://www.senescence.info/aging_theories.html
16. https://www.imminst.org/cureaging/
17. https://www.ncbi.nlm.nih.gov/pmc/articles/PMC4410392/
18. https://www.ncbi.nlm.nih.gov/pmc/articles/PMC2995895/
19. https://pubmed.ncbi.nlm.nih.gov/11976218/
20. https://www.amazon.com/Ending-Aging-Rejuvenation-Breakthroughs-Lifetime/dp/0312367066/
21. https://www.technologyreview.com/s/404453/the-sens-challenge/
22. https://web.archive.org/web/20130606111748/http://www.mprize.com/index.php?pagename=newsdetaildisplay&ID=0104
23. https://www.smithsonianmag.com/innovation/human-mortality-hacked-life-extension-180963241/
24. https://www.sens.org/our-research/intro-to-sens-research/
25. https://www.bbvaopenmind.com/wp-content/uploads/2017/01/BBVA-OpenMind-Undoing-Aging-with-Molecular-and-Cellular-Damage-Repair-Aubrey-De-Grey.pdf
26. https://web.archive.org/web/20170830105451/https://www.leafscience.org/sens-where-are-we-now/
27. https://www.levf.org/
28. https://www.levf.org/projects/robust-mouse-rejuvenation-study-1
29. http://www.cell.com/cell/fulltext/S0092-8674(13)00645-4
30. https://www.ncbi.nlm.nih.gov/pmc/articles/PMC4852871/
31. http://youtu.be/xI38YRz1bbQ
32. https://www.buckinstitute.org/news/leading-scientists-identify-research-strategy-for-highly-intertwined-pillars-of-aging/
33. https://www.libertaddigital.com/ciencia-tecnologia/ciencia/2018-01-19/gines-morata-el-ser-humano-podra-llegar-a-vivir-entre-350-y-400-anos-1276612414/
34. https://www.longecity.org/forum/page/index2.html/_/feature/book
35. https://www.aging-us.com/article/204248/text
36. https://www.sciencedirect.com/science/article/pii/S0092867422013770
37. https://longevity.technology/news/a-decade-of-research-expands-the-hallmarks-of-aging-from-nine-to-twelve/
38. https://www.britannica.com/science/aging-life-process
39. https://www.amazon.com/Handbook-Biology-Aging-Eighth-Handbooks/dp/0124115969
40. https://cdn.who.int/media/docs/default-source/classification/icd/historyoficd.pdf

41. https://www.who.int/news-room/detail/18-06-2018-who-releases-new-international-classification-of-diseases-(icd-11)
42. http://www.who.int/about/what-we-do/gpw-thirteen-consultation/en/#
43. https://www.ncbi.nlm.nih.gov/pmc/articles/PMC4471741/
44. https://www.frontiersin.org/articles/10.3389/fgene.2015.00326/full
45. https://www.amazon.com/Lifespan-Why-Age_and-Dont-Have/dp/1501191977/

3

The World's Biggest Industry?

And that's why the budget I send this Congress on Monday will include a new Precision Medicine Initiative that brings America closer to curing diseases like cancer and diabetes, and gives all of us access, potentially, to the personalized information that we need to keep ourselves and our families healthier.
Barack Obama, 2015

With the science now catching up to the aspirations of 'life-extensionists', this is truly the biggest money fountain we have ever seen.
We are on the verge of a lifespan revolution. Life expectancy is going to rise to between 110 and 130 in the next 30 years. This is not science fiction.
Jim Mellon, 2017

You could make a pill that added two years to a person's life that would be a $100 billion company.
Sam Altman, 2018

New industries have appeared throughout the history of humanity thanks to technologies that were considered impossible until they became realities. Many of these industries were totally discredited by the "experts" of their time. Fortunately, these businesses grew rapidly to become fundamental parts of the global economy.

Many of the most important industries of our world today were once ridiculed. Several important technologies and industries went from impossible to essential. Consider the following inventions and discoveries as examples:

© The Author(s), under exclusive license to Springer Nature Switzerland AG 2023
J. Cordeiro, D. Wood, *The Death of Death*, Copernicus Books,
https://doi.org/10.1007/978-3-031-28927-9_3

1. Trains.
2. Telephones.
3. Cars.
4. Airplanes.
5. Atomic energy.
6. Space flights.
7. Personal computers.
8. Mobile phones.

From "Impossible" to "Indispensable"

The world changes and we change with it. Let us reflect briefly on the beginnings of each of the above industries and what some of the "experts" of their time said:

1. **Trains** were inconceivable to many people, as for centuries humans moved mainly on foot, although the upper classes of some societies could have access to horses on land and boats in the water. During the first half of the 19th century in England some pioneers started the development of trains. Until then, the fastest method of land transport used by the richest and most powerful were horses and carriages. The English publication *The Quarterly Review* wrote the following in 1825:[1]

   What can be more palpably absurd than the prospect held out of locomotives traveling twice as fast as stagecoaches?

2. **Phones** were inconceivable to most of the people until Scottish inventor Alexander Graham Bell began his experiments in Boston in the second half of the 19th century. However, many people still thought they were impossible or infeasible, as shown by the following comments in 1876 by Western Union, then the world's largest telegraph company, and Sir William Preece, chief engineer of the British Post Office, respectively:[2]

   This 'telephone' has too many shortcomings to be seriously considered as a means of communication. The device is inherently of no value to us.
   The Americans have need of the telephone, but we do not. We have plenty of messenger boys.

3. **Commercial cars** appeared in Europe and the United States during the first half of the 20th century but were considered products for the rich

when they were invented. Until the American businessman Henry Ford mass-produced the industrial system with assembly lines, technicians specialized in different parts of the process assembled the cars almost one by one.

The creation of the famous Ford Model T (called, with great irony, the most "popular" car of any color, as long as it was black) allowed the volume of cars to increase, with the consequent reduction of prices and the democratization of access to vehicles. However, Ford's view is summed up in the saying attributed to him:[3]

If I had asked people what they wanted, they would have said faster horses.

4. **Airplanes** were also impossible until they were possible. There are plenty of comments from "experts" at the time who explained why flying was impossible, from comprehensive articles in the prestigious *The New York Times* to statements by the most prestigious scientists of the time. For instance, Scottish physicist and mathematician William Thomson, known as Lord Kelvin, apparently said the following in 1902:[4]

No aeroplane will ever be practically successful. Neither the balloon, nor the aeroplane, nor the gliding machine will be a practical success.

Those statements were preceded by these others in 1896, as a former president of the prestigious *Royal Society of London*, when he ratified his "scientific" belief that planes were inconceivable:[5]

I have not the smallest molecule of faith in aerial navigation other than ballooning… I would not care to be a member of the Aeronautical Society.

Fortunately, the American brothers Orville and Wilbur Wright, who completed only three years of high school, ignored these "scientific" comments, and managed to fly for the first time in 1903. Although the first flight was only for a few seconds, the rest is history.

5. **Atomic energy** was considered scientifically impossible until the first half of the 20th century. In fact, the very word "atom" means precisely indivisible (in Greek: "atomos", which means "indivisible" with "no parts"). The American physicist Robert Andrews Millikan, winner of the Nobel Prize in Physics in 1923, said in the magazine *Popular Science* in 1930:[6]

No "scientific bad boy" ever will be able to blow up the world by releasing atomic energy.

The German physicist Albert Einstein, winner of the Nobel Prize in Physics in 1921, also mistakenly predicted in 1932:[7]

There is not the slightest indication that nuclear energy will ever be obtainable. It would mean that the atom would have to be shattered at will.

The first nuclear fission experiments were conducted in Germany in 1938, demonstrating that the two Nobel laureates, and many other scientists, were wrong. However, it was in the United States that the first atomic bombs were developed through the then-secret Manhattan Project in 1945. Weapons that changed the course of history and put an end to World War II in the Pacific.

6. **Space flights** were perhaps even more "impossible" than planes and atomic energy put together. No one had flown yet at the beginning of the last century on planet Earth itself, so getting out of the atmosphere seemed even more incredible. In the first half of the twentieth century several groups of people, especially in Germany, the United States and Russia, from beginners to scientists, devoted themselves to the task of achieving the unthinkable. Critics, however, kept attacking the "madness" of flying into space, as shown by an editorial in The New York Times in 1920:[8]

   [On the ideas of Robert H. Goddard, the rocket pioneer] that professor does not know the relation of action to reaction, and of the need to have something better than a vacuum against which to react – to say that would be absurd.

   Years after the conclusion of World War II, and in the midst of the Cold War, the Soviets managed to launch Sputnik, the first artificial satellite, in 1957, followed by the first orbital flight in 1961, manned by Soviet cosmonaut Yuri Gagarin. Six weeks later, the American president of the moment, John Fitzgerald Kennedy, announced that the United States would put the first man in the Moon before the end of the decade. Although it had seemed impossible, since the ignorance of both science and technology about travelling in space was enormous, the American astronaut Neil Armstrong became the first human to step on the Moon just 8 years later. Then he said the unforgettable phrase that many of us could see live when we were younger:[9]

   That's one small step for a man, one giant leap for mankind.

7. **Personal computers** have been another technology that has developed exponentially in the 20th century since their humble beginnings as descendants of the first abacuses invented in Mesopotamia 5,000 years ago. American businessman Thomas Watson, president of IBM (International Business Machines), reportedly said in 1943:[10]

I think there is a world market for maybe five computers.

Although Watson may not actually have said that, the reality is that computers were then stunningly large, expensive, and heavy machines. It was hard to appreciate how much smaller, cheaper, and lighter they would become in the future. This is what *Popular Mechanics* magazine said in 1949 when it commented on the first large ENIAC computer in the United States:[11]

Where a calculator like ENIAC today is equipped with 18,000 vacuum tubes and weighs 30 tons, computers in the future may have only 1000 vacuum tubes and perhaps weigh only 1½ tons.

Computers were not conceived for individual use, and the concept of personal computers was difficult to conceive even for entrepreneurs like American engineer Ken Olsen, co-founder and president of DEC (*Digital Equipment Corporation*), who said publicly in 1977:

There is no reason for any individual to have a computer in his home.

Fortunately, thanks to Moore's Law, named in honor of American scientist and businessman Gordon Moore, we have seen computers doubling their power every two years or less, over six decades, while their price continues to fall.

8. **Mobile phones** were born thanks to the convergence of several previous technologies, such as fixed telephones, radio, and personal computers. Although they were almost inconceivable at the time, today almost everyone has a mobile phone if they so wish. Ranging from children to "the elderly", people today have mobile phones, from very cheap models produced in China and India for the equivalent of just ten dollars, to more sophisticated models for about a thousand dollars.

But mobile phones are no longer simple "stupid" phones. In just a decade, phones have become "smart". However, as recently as 2007, when Apple's iPhone appeared and helped popularize smart phones, U.S. businessman Steve Ballmer, then president of Microsoft, said at a conference, according to the newspaper USA Today:[12]

There's no chance that the iPhone is going to get any significant market share. No chance.

Thanks again to Moore's Law, new versions of mobile phones are getting smarter. Today's mobile phones do a huge number of things, and phone calls are only a small part of their use. Thanks to all the new applications, devices and sensors, the new mobile phones have functions ranging from powerful cameras to advanced medical assistants. Within a few years, with new smartphones continuously connected to the high-speed Internet for free or nearly free, there will be no limits to human knowledge. We are advancing rapidly toward the democratization of all the accumulated wisdom from the beginning of civilization. These impressive advances will have implications of every kind, from communications to medicine, as the BBC indicates in a futuristic article that mentions multiple possibilities:[13]

> It's a summer morning in 2040. The internet is all around you and all the things that you're about to do during your day will fall in to place thanks to the data streams flying across the internet. Public transport to the city dynamically adjusts schedules and routes to account for delays. Buying your kids the perfect birthday presents is easy because their data tells your shopping service exactly what they will want. Best of all, you're alive despite a near-fatal accident last month because doctors in the hospital's emergency department had easy access to your medical history.

Today the majority of people consider all these industries as essential fundamental parts of today's civilization, although there are groups that do not use them or even do not want them, because they live in other times with other ideas. For example, many of the Amish communities in North America and the Yanomami aborigines in South America do not want to use these technologies. They prefer to live in their past worlds, like other traditional communities in Papua New Guinea and other parts of the world. These groups have the right to live in the world they want, but they cannot impose their ideas on others. Nor can they stop the scientific advances that arise from the innate curiosity of *Homo sapiens sapiens* since before our departure from the African lands in which we evolved millions of years ago.

A New Industry is Born "Impossible" and will Become Indispensable Very Soon

We have already seen how many "experts" have been wrong throughout history about the development of trains, telephones, cars, airplanes, atomic energy, space flight, personal computers, and mobile phones. There are countless other examples we could cite: radio, television, robots, artificial intelligence, quantum computing, nanomedicine, molecular assemblers, space bases, nuclear fusion, *hyperloop*, brain-computer interfaces, non-animal cultured meat, organ transplants, artificial hearts, therapeutic cloning, cryopreservation of cells and tissues, organ bioprinting, and a huge list of technologies that have been developed since the beginning of the 21st century. The most fascinating, and precisely the subject of this chapter, is the birth of the human rejuvenation industry.

Since the beginning of this century, thanks to scientific advances that allow us to better understand the processes of aging and anti-aging, an industry that until the twentieth century was scientifically "impossible" is emerging, and in the first half of the twenty-first century may finally become reality. We are talking about the human rejuvenation industry, which has the potential to become the largest industry in history, since the great enemy of all humanity is aging. Aging-related diseases cause the greatest suffering to the greatest number of people, especially in advanced countries, where about 90% of the population succumbs to the horror of aging. This was the sad reality until today, when we already have reliable evidence that both the control of aging and rejuvenation are possible. Proofs of concept already exist in cells, in tissues, in organs and in model organisms such as yeasts, worms, flies and mice.

We live in a historic moment in which we have for the first time the scientific opportunity and the ethical responsibility to end humanity's greatest tragedy. Today we know that curing aging is possible; we also know that it will not be easy, we still have much to learn and discover, and we will have to invest enormous amounts of resources of all kinds (human, scientific, financial, etc.). Despite all the future problems, many of them still unforeseen and even unpredictable, today we can finally see that there is a light at the end of the tunnel.

British entrepreneurs Jim Mellon and Al Chalabi published a visionary book in 2017 called *Juvenescence: Investing in the Age of Longevity*. In this book, the authors forecast that average life expectancy will increase to 110–120 years over the next two decades and increase rapidly thereafter. The old paradigm of birth, study, work, retirement and death will be replaced by long lives where we will continually reinvent ourselves in time, as the book's own website states:[14]

So, in summary, the book does three things: firstly, describes the current or soon to be marketed treatments that will enable everyone to live much longer than actuarial table currently suggest. Secondly, it outlines the technologies that have the potential for extending life, such as genetic engineering and stem cell therapy. And lastly, Jim and Al have carefully curated three portfolios for interested investors to consider.

Mellon and Chalabi begin their book with a preface entitled "Longevity Takes Flight", where they compare the aviation industry of a century ago and the rejuvenation industry of today:

> Just as with aviation a century ago, anti-aging science is about to take flight…
>
> It has only been slightly over 100 years since Mr. Boeing built his first aircraft, a seaplane with two seats, and only about 120 years since the Wright Brothers made history with their first flight at Kitty Hawk. Imagine being alive in 1915: could any one of us have ever believed what an aircraft would look like just one short century hence? Almost certainly not. But the really important thing was that by 1915 the mechanism by which aircraft could fly had been discovered – and from that point onwards the design and capability of machines that were able to fly could only improve.
>
> Knowledge, once learned, cannot be unlearned, and despite the occasional interruptions to fundamental human progress (war, famine and plague), it is simply wonderful that today we sit on such a huge repository of information – knowledge which is doubling, in quantity if not quality, every second year. Admittedly a lot of this "knowledge" is not very useful, but there is no doubt that the internet has facilitated a huge improvement in the transmission and use of scientific data -and this for the benefit of all of humanity.
>
> That same pattern of accumulated knowledge applied to aviation is also manifested in the field of aging and longevity. Until the Second World War, aging was a marginal science at best, and this was because very few people beyond the realms of science fiction could envision many human beings living much beyond 100.
>
> Scientists now have a good understanding of the basic genetic makeup of humans thanks to the unveiling of the human genome at the turn of the 21st century and the discovery of the structure of DNA some 50 years prior. Researchers into aging are now wrestling with two key issues:
>
> 1. How to cure or tame the diseases that become more prevalent and devastating as people age; and
> 2. How to research aging as a unitary disease in itself -or putting it another way, as a type of disease state.
>
> The fundamental ways in which our cells work are now being examined with a view to understanding how we might slow, stop or even reverse the process of aging. There are multiple pathways implicated in aging, and the science of discovering and altering them is still in its infancy, but it's an area undergoing explosive growth.

An Ecosystem for the Scientific Rejuvenation Industry Appears

The scientific industry of anti-aging and rejuvenation has barely begun. Unfortunately, there has long been another pseudoscientific industry that endures and survives for decades, centuries, millennia and even before. Miracle potions, fantastic pills, amazing lotions, magical creams, spiritual invocations and supernatural prayers have existed since time immemorial and are likely to continue to exist for many years to come. However, thanks to exponential technological progress, we hope that the light of science will push back the darkness of pseudoscience.

That is why it is fundamental to support the work of scientists who are working hard to achieve humanity's first great dream, as we saw before: immortality (or more rigorously: amortality). Although they may not say so many times, the basic idea is to defeat, both scientifically and morally, the great enemy of all humanity – aging – by far the biggest cause of humanity's suffering today.

One of the most renowned scientists on these subjects is the aforementioned American geneticist, molecular engineer and chemist George Church, professor of genetics at Harvard Medical School, and professor of Health Sciences and Technology at Harvard and at MIT. Church has been involved in such important work as the Human Genome Project, and the Brain Research through Advancing Innovative Neurotechnologies (BRAIN) project, to understand how the human brain is connected and how it works, as well as many other projects, such as copying genes from extinct mammoths into the Asian elephant genome. Church is working on animal rejuvenation, including dog testing at Rejuvenate Bio in order to learn and then apply his findings to humans.[15] Among many other things about anti-aging, Church has also said in the Washington Post that thanks to the expected advances in gene therapies such as CRISPR and other technologies:[16]

A scenario is, everyone takes gene therapy — not just curing rare diseases like cystic fibrosis, but diseases that everyone has, like aging.

One of our biggest economic disasters right now is our aging population. If we eliminate retirement, then it buys us a couple of decades to straighten out the economies of the world. If all those gray hairs could go back to work and feel healthy and young, then we've averted one of the greatest economic disasters in history.

Someone younger at heart should replace you, and that should be you. I'm willing to. I'm willing to become younger. I try to reinvent myself every few years anyway.

Church is co-founder, shareholder and advisor to many companies including Veritas Genetics (genome studies), Nebula Genomics (genome sequencing), Warp Drive Bio (natural products), Alacris (cancer systems therapeutics), Pathogenica (virus and microbial diagnostics), AbVitro (immunology), Gen9 Bio (synthetic biology), Rejuvenate Bio (animal rejuvenation), EnEvolv (genetic engineering), among others. He is also the author, together with the science writer Edward Regis, of the book *Regenesis* where they propose a new genesis for *Homo Sapiens 2.0*. The subtitle of the book is *How Synthetic Biology Will Reinvent Nature and Ourselves*. In addition, the book is also written in DNA, the first book so written in the world, encoded in a small bottle that is delivered along with the printed book. The book ends with a debate about the possibility of future biological immortality, the new technological evolution beyond the old biological evolution, and what Church calls the "end of the beginning" with the beginning of transhumanism (when humans transcend our limitations thanks to science and technology).[17]

Another well-known American scientist immersed in these subjects is the American biochemist, geneticist, and businessman Craig Venter, founding president of Celera Genomics, who became world famous when he started his own Human Genome Project in 1999 outside the public budget, using more advanced technologies that allowed him to finish the genome sequence much faster and cheaper. Venter is also famous for creating the first artificial bacterium in 2010 after rewriting the genome of a bacterium and modifying it to create an artificial life form: "the first self-replicating species we've had on the planet whose parent is a computer" he explained at the time. This synthetic bacterium would later be named Synthia, to commemorate that its genome had been artificially reconstructed in the laboratory.

Venter also co-founded Human Longevity Inc (HLI) in 2014, with the goal to extend healthy living by analyzing individuals' genomes and other medical data, supported by artificial intelligence and deep learning techniques. HLI co-founder Peter Diamandis, a U.S. physician and engineer from Harvard and MIT, as well as co-founder with Ray Kurzweil of Singularity University, has said that, thanks to technological advances, we are going to radically extend our lives and soon "we don't have to die".[18] HLI's mission envisions:[19]

Aging is the single biggest risk factor for virtually every significant human disease… our goal is to extend and enhance the healthy, high-performance lifespan and change the face of aging. For the first time, the power of human genomics, informatics, next generation DNA sequencing technologies, and stem cell advances are being harnessed in one company, Human Longevity Inc., with the leading pioneers in these fields. Our goal is to solve the diseases of aging by changing the way medicine is practiced.

Venter's team also succeeded in 2016 in synthesizing a bacterial genome with the lowest possible gene expression, only 473 genes. This is the first form of life created entirely by humans, baptized as *Mycoplasma laboratorium* to indicate its design in a laboratory. It is hoped that this line of research will lead to the development of bacteria manipulated to generate specific reactions that allow, for example, the production of drugs or fuels. The creation of personalized medicines is also expected thanks to this and other advances in synthetic biology.

Venter has written two books, the first on the sequence of the human genome, specifically his own genome, and the second with the title *Life at the Speed of Light: From the Double Helix to the Dawn of Digital Life*, which deals with the new horizon of scientific frontiers. The book offers an opportunity to reflect again on the old question "What is life?" and to examine what it really means to "play God" from the point of view of the first person to have created artificial life. Venter is a visionary at the dawn of a new era of genetic engineering and the opportunities that emerge from the digitization of life itself.[20]

Another aging expert is American molecular biologist and biogerontologist Cynthia Kenyon, originally known for her genetic studies to understand aging in the tiny worm nematode *Caenorhabditis elegans* (better known as *C. elegans*), one of the most widely used model organisms in biology. She is currently Vice President of Aging Research at Calico (California Life Company, founded by Google in 2013), where this review is found:[21]

In 1993, Kenyon's pioneering discovery that a single-gene mutation could double the lifespan of healthy, fertile *C. elegans* roundworms sparked an intensive study of the molecular biology of aging. Her findings showed that, contrary to popular belief, aging does not "just happen" in a completely haphazard way. Instead, the rate of aging is subject to genetic control: Animals (and likely people) contain regulatory proteins that affect aging by coordinating diverse collections of downstream genes that together protect and repair the cells and tissues. Kenyon's findings have led to the realization that a universal hormone-signaling pathway influences the rate of aging in many species, including humans. She has identified many longevity genes and pathways, and her lab was the first to discover that neurons, and also the germ cells, can control the lifespan of the whole animal.

Kenyon has made powerful statements based on her research, and has even raised the possibility of biological immortality, as she explains to the San Francisco Gate interviewer:[22]

> In principle, if you understood the mechanisms of keeping things repaired, you could keep things going indefinitely.
>   I think that [immortality] might be possible. I'll tell you why. You can think about the life span of a cell being the integral of two vectors in a sense, the force of destruction and the force of prevention, maintenance, and repair. In most animals the force of destruction has still got the edge. But why not bump up the genes just a little bit, the maintenance genes. All you have to do is have the maintenance level a little higher. It doesn't have to be much higher. It just has to be a little higher, so that it counterbalances the force of destruction. And don't forget, the germ lineage is immortal. So it's possible at least in principle.

In an article written by Kenyon under the title "Aging: The Final Frontier," she explains that:[23]

> One might think that many genes would have to be changed in order to extend our lifespans – genes affecting muscle strength, wrinkles, dementia, and so forth. But research on worms and mice has found something quite surprising: there are certain genes whose alteration can slow the aging of the whole animal all at once.

Scientists Church, Venter and Kenyon are examples of a generation that is working openly on anti-aging and rejuvenation issues in prestigious institutions without fear of saying so publicly. Behind them there is a new generation of scientists following in their footsteps, such as the aforementioned Portuguese microbiologist João Pedro de Magalhães.[24]

Among his many scientific research projects related to longevity, de Magalhães has sequenced and analyzed the genome of the Greenland whale. He has also contributed to the analysis of the naked mole rat genome. Both mammals are exceptionally long-lived and resistant to cancer. On his website, he writes a few words that hopefully will motivate others:[25]

I hope senescence.info can also make people aware of the problem that is aging. Aging will likely kill you and those you love. It is the main reason why great artists, scientists, sportsmen, and thinkers die. Our society and religion make it easier to accept aging and the inevitability of death. I believe that if people thought more about death and how horrible it is, a greater effort would be made to avoid death and invest more in biomedical research and in understanding aging in particular.

In an even younger generation we find the New Zealand born American scientist and investor Laura Deming. Born in 1994 in New Zealand, she was educated by her parents at home. At the age of 8 she became interested in the subject of aging and at the age of 12 she began to work as an intern at Cynthia Kenyon's laboratory in San Francisco. At the age of 14 she was accepted to study at MIT, but in 2011 she dropped out and returned to California as one of the first so-called *Thiel Fellows* – beneficiaries of $100,000 funding from Peter Thiel for leaving university and starting a business.[26]

Deming is a partner and founder of *The Longevity Fund*, a venture capital firm focused on aging and life extension. She believes science can be used to achieve biological immortality in humans, and has said ending aging "is a lot closer than you might think." Her company's website says so:[27]

Investing in Human Longevity

In the 20th century, we learned that healthy lifespan is malleable. The scientific pathways behind this effect are incredibly complex and difficult to correctly control, but manipulating them may lead to new treatments for age-related disease. We want to get these therapeutic advances translated safely to patients as quickly as possible.

Longevity Fund companies have collected >$500M in follow-on funding, resulting in the 2018 IPO of the first company dedicated to reversing the diseases of aging, and multiple programs in clinic to reverse or prevent age-related disease.

De Magalhães and Deming are excellent examples of two younger researchers who have left behind the stigma of talking about, and even more researching, anti-aging and rejuvenation, taboo subjects that used to destroy the careers or scientific credibility of some researchers, who could be confused with supporters of the pseudo-science of magical anti-aging or miraculous rejuvenation.

Science and Scientists, Attracting Investments and Investors

Scientific advances in recent decades have begun to attract investors to fund more research. Now that science and scientists have begun to obtain real results, even if it is still at the level of organisms such as worms or mice, we could say that the die is cast, or as the Roman Julius Caesar said when crossing the Rubicon River: *alea iacta est!*

Alongside public investments, we now have the possibility of seeking private investment to carry out more research which, we expect, will soon generate more human clinical trials based on positive results in animals. Thus is born the industry of anti-aging and scientific rejuvenation, with the potential to become the largest sector of the economy and to transform the history of humanity: before and after the inevitability of death.

Moldavian engineer and entrepreneur Dmitry Kaminskiy leads the global initiative www.Longevity.International, which aims to accelerate the development of rejuvenation technologies for its application and commercialization. The platform self describes as:[28]

> Longevity International is an open-source non-profit decentralized knowledge and collaboration platform, to promote more efficient and intelligent collaboration, cooperation, and productive discussion among Longevity Industry players, to accelerate and democratize Longevity industry collaboration and harmonization.

After joining forces with other organisations (including three based in the UK: Aging Analytics Agency, Biogerontology Research Foundation and Deep Knowledge Life Sciences), they have published an impressive series of reports that have grown, improved, and increased over time. These reports contain lots of material useful for anyone who wants to understand the development of the industry since their publications began in 2013:[29]

2013: *Regenerative Medicine: Industry Framework* (150 pages).

2014: *Regenerative Medicine: Analysis & Market Outlook* (200 pages).

2015: *Big Data in Aging & Age-Related Diseases* (200 pages).

2015: *Stem Cell Market: Analytical Report* (200 pages).

2016: *Longevity Industry Landscape Overview* (200 pages).

2017: *Longevity Industry Analytical Report 1: The Business of Longevity* (400 pages).

2017: *Longevity Industry Analytical Report 2: The Science of Longevity* (500 pages).

2018: *Longevity Industry Landscape Overview. Volume I: The Science of Longevity* (701 pages).

2018: *Longevity Industry Landscape Overview. Volume 2: The Business of Longevity* (650 pages).

The first 2017 report deals with the longevity business, and its executive summary begins and ends like this:[30]

Biotechnology and geroscience in particular are on the verge of a Cambrian explosion of breakthrough science that will transform healthcare into an information science capable of improving the human condition more profoundly than even the advent of antibiotics, modern molecular pharmacology, and the Green agricultural revolution. The time-course of this major evolutionary transition and whether we and our loved ones live long enough to benefit from these breakthroughs is dependent upon the choices of the scientific and investment community today.

The report presents a comprehensive overview of the longevity situation at the enterprise level, including a comprehensive analysis of large public and private companies, new start-ups working on anti-aging and rejuvenation technologies, research centers, foundations, and universities. The interactive system of www.Longevity.International allows multiple types of analysis, for example: monitoring international investment flows, studying links between scientists and investors, using Big Data in the growing international database, generating networks with specific interests, creating maps of connections between different institutions and making visualizations of groups and "clusters". Figure 3.1 shows a cluster analysis among a group of more than 100 companies from different countries dedicated to anti-aging and rejuvenation.

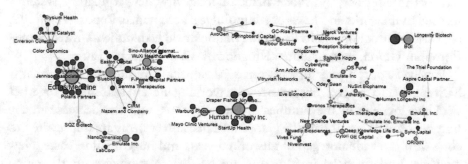

Fig. 3.1 Cluster analysis among different companies. (Source: Longevity.Iinternational 2020; http://www.Longevity.International)

It is also possible to obtain a panoramic view of the general scientific landscape and of potential businesses related to it. The report includes lists of leading scientists (such as some of those named above: Nir Barzilai, George Church, Aubrey de Grey, João Pedro de Magalhães, Cynthia Kenyon, David Sinclair, etc.), major investors (Jeff Bezos, Michael Greve, Jim Mellon, Peter Thiel, Yuri Milner, Sergey Young, etc.) and major *influencers* (such as Sergey Brin, Larry Ellison, Ray Kurzweil, Larry Page, Craig Venter, etc.). Also included are lists of conferences, books, publications, and events related to "geroscience" in general. It is worth mentioning that both the World Economic Forum in Davos and the British publication *The Economist* have begun to organize events on the subject of aging and on the possibilities of anti-aging to solve the serious economic crisis that is approaching due to the accelerated aging of the population, a fact that is widespread even in many poor countries.

Longevity International's first report also indicates the astronomical costs of attacking the consequences rather than the causes of aging. For example, the costs of cancer treatments are around $900 billion a year worldwide, followed by more than $800 billion for dementia, nearly $500 billion for cardiovascular disease, and hundreds of billions more for other age-related diseases. Health systems have shifted from health care to disease care, and basically diseases of aging, as the report explains.

Continuing the work of Longevity International in 2018, three additional reports were published:

2018: *Longevity Industry Analytical Report 3: 10 Special Cases.*

2018: *Longevity Industry Analytical Report 4: Regional Cases.*

2018: *Longevity Industry Analytical Report 5: Novel Financial Instruments.*

Report 3 focuses on cases of regenerative medicine, gene therapies, aging biomarkers, stem cell treatments, geroprotective and nutraceutical products, artificial intelligence and blockchain for longevity, new regulatory systems, sectoral frameworks and levels of technological acceptance. Report 4 compiles regional information, basically on the major world economies: United States, European Union, Japan, United Kingdom, Asia and Eastern Europe.

Finally, Report 5 puts forward financial solutions to the aging crisis that is looming due to the growing gap between working life and retirement, together with the increase in the number of retired people and the decrease in the number of workers. Today's insurance and pension systems are unprepared to cope with the widening gap of unworked years and huge medical costs. New strategies and financial instruments are needed to capitalise on the aging industry, close the tax gap and rejuvenate human beings. New schemes such as venture capital funds, hedge funds and trust funds are proposed for the coming years with the aim of financing rejuvenation and transition to a new economy.

Longevity International shows a route or a way to pass from a world where there is aging to the world of the future, where rejuvenation will prevail. Soon we will have many more ideas, as this industry is just beginning. But the important thing is to start and prepare the way for the radical extension of lifespans and healthspans. The following more recent reports by Longevity International, with hundreds of pages full of useful information of this rapidly growing industry, were also led by Kaminskiy and his team at Deep Knowledge Group:

2020: *Longevity Industry 1.0: Defining the Biggest and Most Complex Industry in Human History.*
2021: *Biomarkers of Human Longevity*

All these reports contain a vast wealth of data, much of which will eventually benefit from artificial intelligence to process it faster, cheaper, and better. The Longevity International web platform keeps accumulating valuable information that is continuously updated to monitor the global longevity ecosystem. The website allows the analysis of the data and also create "mind maps" with over 20,000 companies, more than 9,500 investors and over 1,000 R&D centers identified until 2022. This shows an incredible growth of an ecosystem which barely existed one decade ago. Now it is possible to search the information by countries, regions, companies, revenues, funding, investors, government agencies, R&D centers, technology sectors, finance sectors, personalities, etc. Figure 3.2 shows such a country classification with the over 20,000 companies by country in 2022.[31]

According to the *Longevity Marketcap Newsletter*, a web publication launched in July 2020 to monitor the impressive growth of the longevity industry, there was about $3 billion total funding in 2021 based on publicly announced financings, with 36 completed capital raises (including 32 private venture deals, 2 IPOs, 1 SPAC and 1 acquisition). This is truly an impressive growth in a decade with five $100+ million new venture announcements in 2021:[32]

In 2011, Laura Deming founded the first longevity venture capital fund with just $4M raised. Ten years later, five venture funds or venture builders each announced new $100M+ funds and commitments in 2021 alone. Including:
• Kizoo Technology Capital – Michael Greve announced commitment of $340M USD toward rejuvenation biotech startups.
• Apollo Health Ventures – Announced the close of $180M for their second fund.

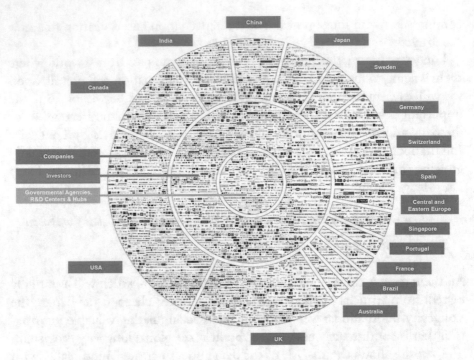

Fig. 3.2: Exponential growth of the longevity ecosystem around the globe. (Source: Longevity.International 2022; https://www.longevity.international/longevity-ecosystem-by-country)

- Korify Capital – based in Switzerland, announced new $100M+ longevity biotech venture fund.
- Maximon – Announced a $105M venture building fund.
- Cambrian Biopharma – Technically not a venture fund, but operates as a holding company that does venture creation, essentially. Announced a $100 Series C in 2021.

Also worth mentioning is VitaDAO raising $5M in ETH – a major milestone as a new decentralized structure enabled by web3 to fund and commercialize longevity research IP.

Billions of dollars are pouring into longevity, including the 2021 financing behind Altos Labs with Jeff Bezos and Yuri Milner, among others committing $3 billion, and also NewLimit financed by billionaire crypto investor Brian Armstrong, founder and CEO of the exchange platform Coinbase. These investments also indicate the arrival of cryptocurrencies to finance the longevity industry, with other million-dollar investments by the young Vitalik Buterin, the Russian-Canadian co-founder of Ethereum, the second largest cryptocurrency after Bitcoin, and many other wealthy "bitcoiners" investing now in health and longevity. In 2022, Saudi Arabia also announced the creation of Hevolution (Health+Evolution) Foundation to finance research on longevity, committing $1 billion per year over the next two decades, and the United Arab Emirates is considering similar initiatives.

The longevity industry has really grown from millions of dollars at the start of this century to billions of dollars today, and surely trillions by the 2030s and 2040s. This will soon be the largest industry in the history of the world, as Michael Greve, German Internet entrepreneur and founder of the Forever Healthy Foundation, explained in an interview with Red Bull in 2021:[33]

I've seen the transformation from no PC to PC, the emergence of the internet, of mobile, of cloud services, of all these things that have completely changed our world. And this is going to be just like the digital revolution, only much, much bigger. Rejuvenation is going to be the most fundamental change we've ever seen in our entire history. To realize that it's also going to be the biggest business in history, all it takes is a bit of math: four billion people over forty, ten dollars a month, makes $480 billion year after year. That's a five trillion-dollar company. That is the magnitude of it, and we're only talking about a single root cause.

At the global level, a newly emerging ecosystem for longevity has already appeared, combining science, finance, business, governments, and other national and international actors. We are in the transition from the local to the global world, where changes are also moving from linear to exponential. Now is the time to try to make this still fragile ecosystem grow exponentially into the world's largest industry, the industry that will lead us to the death of death.

Notes

1. https://books.google.com/books?id=JQ8Gtv4A5tMC&dq=palpably&q=palpably#v=snippet&q=palpably&f=false
2. http://rinkworks.com/said/predictions.shtml
3. https://hbr.org/2011/08/henry-ford-never-said-the-fast
4. http://zapatopi.net/kelvin/papers/interview_aeronautics_and_wireless.html
5. https://en.wikiquote.org/wiki/William_Thomson
6. https://en.wikiquote.org/wiki/Incorrect_predictions
7. http://rinkworks.com/said/predictions.shtml
8. http://www.nytimes.com/2001/11/14/news/150th-anniversary-1851-2001-the-facts-that-got-away.html
9. https://www.youtube.com/watch?v=MypSliQOv2M

10. https://www.pcworld.com/article/155984/worst_tech_predictions.html
11. http://www.popularmechanics.com/technology/a8562/inside-the-future-how-popmech-predicted-the-next-110-years-14831802/
12. https://usatoday30.usatoday.com/money/companies/management/2007-04-29-ballmer-ceo-forum-usat_N.htm
13. http://www.bbc.com/future/story/20141015-will-we-fear-tomorrows-internet
14. https://www.juvenescence-book.com/book-overview/
15. http://www.rejuvenatebio.com/
16. https://www.washingtonpost.com/news/achenblog/wp/2015/12/02/professor-george-church-says-he-can-reverse-the-aging-process/
17. https://www.amazon.com/Regenesis-Synthetic-Biology-Reinvent-Ourselves/dp/0465075703
18. https://www.youtube.com/watch?v=hC3OfWFjdXo
19. https://www.gowinglife.com/the-renaissance-of-rejuvenation-biotechnolog
20. https://www.amazon.com/Life-Speed-Light-Double-Digital/dp/0143125907/
21. https://www.calicolabs.com/people/cynthia-kenyon
22. https://www.sfgate.com/magazine/article/Finding-the-Fountain-of-Youth-Where-will-UCSF-2667274.php
23. https://www.project-syndicate.org/commentary/aging%2D%2Dthe-final-frontier
24. http://genomics.senescence.info/
25. http://www.senescence.info/
26. https://www.fightaging.org/archives/2011/06/a-profile-of-laura-deming-thiel-foundation-fellowship-recipient/
27. https://longevity.vc/
28. https://data.longevity.international/press-release.pdf
29. http://longevity.international/
30. https://www.fightaging.org/archives/2017/10/longevity-industry-whitepapers-from-the-aging-analytics-agency/
31. https://www.longevity.international/longevity-ecosystem-by-country
32. https://sub.longevitymarketcap.com/p/037-jan-11th-2022-longevity-marketcap
33. https://www.redbull.com/int-en/theredbulletin/michael-greve-biohacking-longevity

4

From the Linear to the Exponential World

We tend to overestimate the impact of a new technology in the short run, but we
underestimate it in the long run.
Roy Amara's Law, 1970

By 2029 we will have reached longevity escape velocity.
Ray Kurzweil, 2017

Since the beginning of humanity, science and technology have always been
the main catalysts for change and breakthroughs. Science and technology are
what differentiate the human species from other animal species. Inventions,
creations, and discoveries such as fire, the wheel, agriculture, and writing have
allowed the progress of *Homo sapiens sapiens* from our primeval ancestors in
the African savannas to the first space flights. Thanks to exponential changes
we will soon also be able to control human aging and rejuvenation.

The agricultural revolution was the first great revolution of the human species, around 12,000 years ago. The industrial revolution would come later
after the invention of printing and the scientific development that allowed the
industrialization of societies. Today we are living the third great human revolution, a revolution that has received many names: the intelligence revolution,
the knowledge revolution, the post-industrial revolution, the fourth industrial
revolution, etc.

Futurists such as American engineer Ray Kurzweil, co-founder of Singularity
University and director of engineering at Google, suggest that the world is
moving quickly toward a time when human beings will become much more
advanced thanks to the impressive exponential progress of science and technology. This fundamental transformation has been described as "technological
singularity," perhaps similar to the momentous change in evolution from apes
to humans. We will continue to advance in the extension of life, but also in
the expansion of life.

© The Author(s), under exclusive license to Springer Nature Switzerland AG 2023
J. Cordeiro, D. Wood, *The Death of Death*, Copernicus Books,
https://doi.org/10.1007/978-3-031-28927-9_4

From the Past to the Future

Until the eighteenth century, humanity was constrained by the so-called "Malthusian trap", described by the English cleric and economist Thomas Robert Malthus. In 1798, Malthus published his book *Essay on the Principle of Population*, in which he explained "the perpetual struggle for space and food" and concluded that[1]:

> Population, when unchecked, increases in a geometrical ratio. Subsistence increases only in an arithmetical ratio.
> This implies a strong and constantly operating check on population from the difficulty of subsistence. This difficulty must fall somewhere and must necessarily be severely felt by a large portion of mankind.

His theories are known today as Malthusianism, and in its most modern version as Neo-Malthusianism. Malthusianism is a demographic, economic and socio-political theory developed during the Industrial Revolution, according to which the rate of population growth responds to a geometric progression, while the rate of increase of resources for survival does so in arithmetic progression. For this reason, according to Malthus, in the absence of repressive obstacles such as hunger, wars, and plagues, the birth of new beings would increase the gradual pauperization of the human species and could even provoke its extinction (what has been called Malthusian catastrophe). What Malthus called arithmetic growth is what some today call linear growth, and geometric growth would be the equivalent of current exponential growth.

At the end of the eighteenth century, when Malthus was writing his famous essay, the population of the United Kingdom was not yet 10 million, but he was convinced that there were already too many people at that time and that the country was overpopulated. His ideas made such an impact at that time, as the Industrial Revolution was just beginning, precisely in the United Kingdom, that the British Government decided to carry out the first modern census in its history. The census was completed in 1801 with estimates that there were 8.9 million people in England and Wales, plus 1.6 million in Scotland, a total of 10.5 million in Great Britain. Globally, the world population is estimated to have reached one billion in 1804.

These figures seemed too high to Malthus, and depending on the low technological level of the world at the time, he might have been right in his day. Fortunately, the world has come a long way thanks to the Industrial Revolution and now we can even say that today's poor live better than yesterday's rich, with a lifestyle unthinkable two centuries ago. In addition, average life expectancy has almost tripled from the end of the eighteenth century to the beginning of the twenty-first century. We think that most people today, perhaps with the exception of a few Malthusians, would agree that we are living longer and better than we did then. We owe this precisely to the great advances of humanity, which have allowed us to escape from the Malthusian trap that the English philosopher Thomas Hobbes had also described in his Leviathan in 1651[2]:

> No arts; no letters; no society; and which is worst of all, continual fear, and danger of violent death; and the life of man, solitary, poor, nasty, brutish, and short.

Figure 4.1 shows the sad reality of humanity until the eighteenth century, when economic growth was almost non-existent, measured as per capita income or Gross Domestic Product (GDP) per capita. According to updated figures from the British economic historian Angus Maddison, per capita income was around a thousand dollars a year, the rich had something more, and the poor had something less, but we were all economically poor then, and even worse, we lived a short life. The majority died young, including as children and even at birth, and those who lived longer and overcame the common violent deaths of those times led an existence that today we would consider poor, dirty, brutal, and short, as Hobbes had explained centuries before.

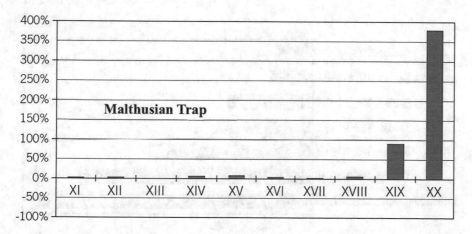

Fig. 4.1 From the Malthusian Trap to the Industrial Revolution. (GDP per capita over the last centuries). (Source: Authors based on Angus Maddison 2007)

The economic growth that began with the industrial revolution has been truly astonishing. The American entrepreneur Peter Diamandis, co-founder of Singularity University and Human Longevity Inc., among many other companies, indicates that we are experiencing exponential changes that are radically transforming the world economy[3]:

In the next 10 years, we will create more wealth than was created in the past century.

In *Abundance: The Future Is Better Than You Think (Exponential Technology Series)*, Diamandis and co-author Steven Kotler explain how we are leaving behind a world of scarcity and entering a world of abundance.[4] In fact, thanks to the acceleration of technological change, we believe that in the next two decades we will see more transformations than in the last two millennia. It is difficult to understand at first glance, so we repeat this fundamental idea: we think that in the next 20 years we are going to live through more technological transformations than in the last 2000 years.

Figure 4.2 shows how economic development processes have accelerated. The first country in human history to systematically double its per capita income was the United Kingdom during the Industrial Revolution, taking 58 years to achieve, between 1780 and 1838. The second country was the United States, which managed to double its income in 47 years, between 1839 and 1887. Then came Japan and made it even faster, in 34 years, between 1885 and 1919. Japan was also the first non-Western country to enter the

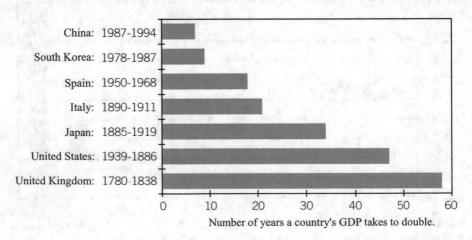

Fig. 4.2 The rapid acceleration of economic growth. (Source: Authors based on Angus Maddison 2007)

developed world, a fact that overturned some colonial prejudices that prevailed at the time, which claimed that only European countries and their more advanced colonies could develop.

China set the world record for economic growth at the end of the twentieth century, when it demonstrated that it is possible to double per capita income in less than a decade. Very positive news for the rest of the world, as other countries are following these examples. Even India has begun to progress at high rates of growth, also followed by countries in Africa and Latin America. Such experiences reveal that there are no longer excuses for countries not to lift themselves out of poverty, which is why the World Bank has set a target to end extreme poverty in the world by 2030.[5] The Sustainable Development Goals (SDG's) of the United Nations (UN) also aim to achieve this goal by 2030. The best thing is that, for the first time in history, there is a real possibility of ending extreme poverty around the world.[6]

Figure 4.3 shows the economic growth of different parts of the world from 1800 to 2016. Until the eighteenth century, average per capita income was only about $1000 per year, worldwide. The industrial revolution changed this tragic situation and created an enormous amount of wealth. The first countries to industrialize were the first to grow, a reality that was maintained for much of the century. Fortunately, the poorest countries are now beginning to grow faster and catch up with those that did before. The vertical axis in Fig. 4.3 indicates an exponential growth in income, which has gone from the historic $1000 up to the eighteenth century to about $10,000 in many countries and

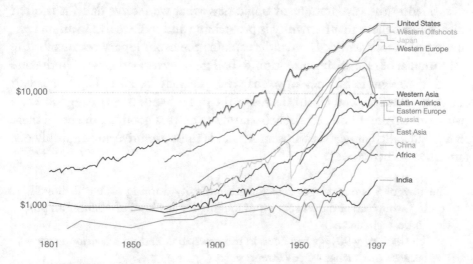

Fig. 4.3 "Exponential" Economic Growth - GDP per capita adjusted for price changes over time (inflation) and price differences between countries (measured in international-$ in 2011 prices). (Source: Authors based on Maddison Project Database 2018)

over $50,000 in the richest countries today. As we move into the twenty-first century, all countries will exceed the income of $10,000 and will probably continue to rise to $100,000 and above. We are moving from scarcity to abundance, although some still don't want to believe it.

Additionally, many prices are also decreasing thanks to our capacity to produce more with less, as described in the book by MIT scientist Andrew McAfee, *More from Less: The Surprising Story of How We Learned to Prosper Using Fewer Resources—and What Happens Next*.[7] Thus, we seem to be advancing to a future of "higher" incomes and "lower" prices.

The Canadian-American psychologist Steven Pinker also explains why we live in "the best time to be alive" because, although it is difficult to believe with the naked eye, we are in the most peaceful times in human history. In his 2012 book *The Better Angels of Our Nature: A History of Violence and Humanity*, Pinker explains that violence has declined worldwide since our first ancestors *Homo sapiens sapiens* appeared in Africa.[8] Pinker continues to demonstrate and defend his theses in his 2018 book *Enlightenment Now*, where he explains the importance of reason, science, and humanism for the progress of humanity.[9]

Towards a Demographic Crisis, But Not the One That Many Fear

Considering the vast amount of tragic news that we receive daily, it is often difficult to believe that humanity is progressing and that we live in an increasingly prosperous world. Diamandis explains the evolutionary reasons for the importance of prioritizing attention to bad news over good news. On the one hand, if we ignore bad news, we may die, precisely because the news is bad and can mean the end of our lives. On the other hand, if we miss good news we are not going to die, precisely because the news is good. In the brain there is a gland called amygdala whose function is to be attentive and multiply our attention to bad news[10]:

> The amygdala is our danger detector. It's our early warning system. It literally combs through all of the sensory input looking for any kind of a danger on putting in on high alert…
>
> Well that's why 90% of the news in the newspaper and on television is negative because that's what we pay attention to…
>
> The media takes advantage of this and you know the old saw, "if it bleeds it leads".

Many think that the world population is growing by leaps and bounds and that this is leading humanity into catastrophe. They are not new ideas; it is more than two centuries ago that Malthus raised them. Today we know that he was wrong, among other things because he did not consider the technological change initiated with the Industrial Revolution. Britain may have been overpopulated with less than 10 million people in the eighteenth century, but that was due to a lack of technology to produce food and other goods and services.

If we go back much further in time, it is estimated that more than 50,000 years ago, no more than one million human beings could survive in Africa, as we lacked technology at the time. With just hunting, fishing and food gathering, the African continent could not support more than a million people. Fortunately, 10,000 years ago, agriculture was invented, which allowed food to be produced and stored so that people could secure their livelihoods. Our ancestors then stopped being nomads in search of food, and founded the first cities with guaranteed food sources. Until the invention of agriculture our ancestors lived in another "Malthusian trap" that, fortunately, was solved thanks to agriculture and other basic technologies that allowed us to reach the eighteenth century.[11]

The issue of world population has always been a vital one, especially for the most powerful nations. At the end of World War II in 1945, the United Nations began making long-term demographic projections. The first projections of a century, up to 2050, showed figures of up to 20 billion people because of the high birth rates in the world at that time. Although the planet's population was in the order of 2.5 billion in 1950, if high growth rates were maintained it would be possible for the world's population to reach 20 billion by 2050. However, birth rates have been falling country by country, and so projections have also been decreasing over the years, from 20 to 18, then 15, then 12, and are now estimated at less than 10 billion in 2050.

The American ecologist Paul Ehrlich wrote in 1968 a world bestseller entitled *The Population Bomb* in which he began by saying[12]:

> The battle to feed all of humanity is over. In the 1970s and 1980s hundreds of millions of people will starve to death in spite of any crash programs embarked upon now. At this late date nothing can prevent a substantial increase in the world death rate...

Fortunately, Ehrlich was absolutely wrong. Hundreds of millions of people did not die in the 1970s, something we owe to continued technological advances such as the so-called green revolution in agriculture, which increased agro-industrial productivity, and thanks to the continued reduction in the birth rate. However, Ehrlich has continued to say and write that the world is overpopulated, waiting for a neo-Malthusian catastrophe, making predictions that always end up proving wrong, while technologies advance and population growth rates continue to fall.

The demographic transitions achieved through the invention of agriculture thousands of years ago, the Industrial Revolution two centuries ago in the face of Malthus' fears, and the Green Revolution decades ago in the face of Ehrlich's fears, put us on the trail of what awaits us with future technologies. In the next few years, we will see the development of biotechnology, nanotechnology, robotics, and artificial intelligence, among other fascinating technologies, which would cause astonishment not only to Malthus and Ehrlich, but also to many of our contemporaries. Nor should we forget that these technological changes are not linear but exponential, increasingly rapid.

The reality today is that the populations of many countries are stabilizing and beginning to decline. For example, the population in China declined by nearly one million people between 2021 and 2022.[13] Figure 4.4 shows the historical demographic evolution from 1950, as well as projections for different parts of the world to 2050. According to the United Nations average

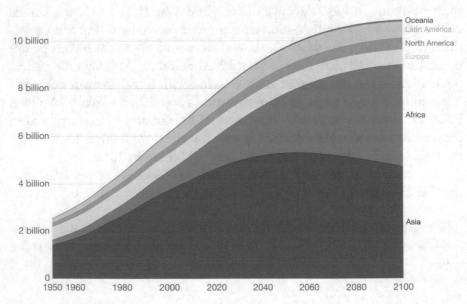

Fig. 4.4 "Linear" population growth - World population projections to 2100 (Projected population to 2100 is based on the UN medium population scenario) (Source: www. ourworldindata.org based on HYDE Database (2016) and United Nations World Populations Prospects 2022)

estimate in 2022, the world will have a population of 9.7 billion in 2050 and 10.4 billion in 2100. These projections also indicate that the world population will reach 9 billion in 2037.[14]

We can already see how the population in different parts of the world is stabilizing and beginning to shrink. This is the case of the population of countries such as Germany, Japan and Russia. If we believe the figures of the United Nations, the population of Japan has already declined from 128.1 million in 2010 to 124.9 million in 2021, and will further decrease to 73.8 million in 2100 according to the average of its estimates. In the most extreme scenario, the Japanese population would fall to just 54.2 million, and a century later perhaps the islands would lose all population if current trends continued. The population is shrinking dramatically because of the lack of births and because many women find it very difficult to conceive because the average age in Japan is now over 40. In short, it is an almost irreversible phenomenon under current conditions. Fortunately, the world will change radically thanks to new technologies, and it is countries like Japan that, for obvious reasons, will be most interested in anti-aging and rejuvenation issues.

In Germany the population is expected to fall from 83.4 million in 2022 to 68.9 million according to average estimates, and 54.6 million in the extreme scenario in 2100. In Russia, the population would decline from 144.7 million in 2022 to 112.1 million in the medium scenario, and 78.6 million in the extreme scenario in 2100. Similar trends can be observed in Catholic countries such as Spain and Italy. In Spain, the population would decrease from 47.6 million in 2022 to 30.9 million according to average projections, and 22.2 million in the extreme scenario in 2100. In Italy, the population would fall from 59.0 million in 2022 to 36.9 million according to average projections, and 26.9 million in the extreme scenario in 2100. So, Italy, the country of the *"bambini"* is left with fewer and fewer children. On the other hand, if the population of countries such as Germany, Spain and Italy do not shrink more or faster, it is due to immigration, which compensates in some way for the reduction in the birth rate of indigenous populations.

Perhaps the most dramatic case in the world will be the large reduction in China's population, partly due to the "one child" policy, which was compulsory for several decades. In China, most of today's citizens are the only children of parents who were themselves only children, something that has created many social distortions. In addition, there are more men than women due to female infanticide in a sexist culture in which fathers prefer to have male descendants. The result is a brutal collapse of the population, something never seen before in human history in peacetime. According to the United Nations, China's population would fall from 1425.9 million in 2022 to 766.7 million according to average projections, and to just 511.3 million in the extreme scenario in 2100. Due to these tragic demographic forecasts, China is another of the countries increasingly interested in anti-aging and rejuvenation.

The coming demographic crisis is no longer an excess of humans, but rather a possible stagnation and reduction of the planet's population. If we analyze the reasons why the world has advanced so much in these two centuries, one of the main reasons is precisely the increase in population. More people thinking, more people working, more people creating, more people innovating, more people discovering, more people inventing. People don't just come into the world with a mouth to eat and a butt to defecate; no, people come into the world with a brain, and the brain is considered to be the most complex structure in the known universe. The brain is a wonderful organ with the ability to imagine and create almost anything.

However, it is true that the population is still increasing in some poor countries in Africa and Asia mainly, but the birth rate is declining and current figures are likely to overestimate what is going to happen, as demonstrated by the historical experience of these projections over the last few decades. On the other hand, the incorporation of humbler people into the global economy will have positive effects, as it is precisely the poorest people who know how to do more with less: they can spur what is known as "frugal innovation".[15] The productivity of the world will increase thanks to the new minds that will be incorporated into the planetary economy. However, even in the poorest countries the population will stabilize over the next few decades.[16]

Figure 4.5 shows the drastic drop in the population aged under 5 and the rapid increase in the population aged over 65 worldwide. These are global trends: the world is aging by leaps and bounds, there are fewer young people and more elderly people. It is a global phenomenon, a problem of both the so-called rich countries and the countries classified as poor. Historically, people used to die from causes related to violence and infectious diseases when they were still young. Now, on the other hand, people die of age-related diseases after long, continuous, terrible personal suffering.

Figure 4.6 shows what used to be called "demographic pyramids", although they are no longer pyramids but rather rectangles. In the most dramatic cases, such as the current Japan and the coming China, the pyramid is being inverted from a triangle with the base down to a triangle with the base up. Figure 4.6 also shows the extent to which the world population is stabilizing and aging rapidly. It is expected that there will be fewer and fewer people in each generation.

World population aging has dramatic human and social implications, as well as serious economic consequences. As age continues to rise, there will be

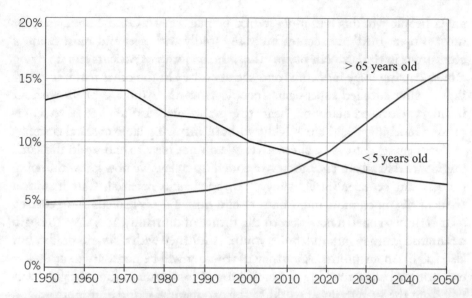

Fig. 4.5 The real demographic crisis (Percentage of the population, %). (Source: Authors based on UN data 2022)

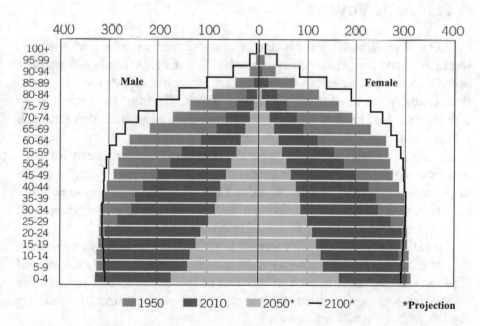

Fig. 4.6 Changing shape of the population pyramids. (Source: Authors based on UN data 2022)

fewer people working and more retired or pensioners. On the one hand, for most people, medical expenses increase rapidly with age, and most medical costs occur in the last years of life. Thus, the majority of patients end up dying after incurring huge individual and social costs. On the other hand, interestingly, patients called super-agers (people who reach the age of 95 without having any cancer, dementia, heart disease, or diabetes) tend to have "compressed morbidity" and die quickly, without incurring large medical costs.[17]

Fortunately, there is an alternative. We do not have to end up in the same tragic way that all our ancestors have ended up today. We now know the aging process can be scientifically slowed, stopped, and reversed. Our historical challenge is much more important: to end humanity's great common enemy.

It is time to open a new path to the future of humanity. It is time to begin a fantastic journey into indefinite youth. A journey with risks, no doubt, but also full of opportunities. A journey in which it will be necessary to cross several bridges before reaching the most longed-for dream of humanity. A journey from the present linear world to the exponential world of tomorrow.

A Fantastic Voyage

In 2004, Ray Kurzweil and his doctor Terry Grossman, a longevity expert, wrote *Fantastic Voyage: Live Long Enough to Live Forever*. The book's name is based on a famous American science fiction film produced in 1966 by twentieth Century Fox starring actress Raquel Welch. The film is a fantastic story of a journey into the human body with a manned submarine that has been reduced in size in a miniaturization center.

The film won two Oscars and inspired a novel of the same name by Russian-American writer Isaac Asimov, a series of cartoons and even a painting of the same name by Spanish painter Salvador Dalí. The American producer James Cameron and the Mexican Guillermo del Toro have shown their interest in making a new version of the film for Hollywood.

Fantastic Voyage is Kurzweil's second book on health. In his first book on the subject, *The 10% Solution for a Healthy Life*, written in 1993, Kurzweil explains how he cured himself of diabetes at the age of 45 and reduced the calorie, fat, and sugar content of his diet, among other changes, to minimize the risk of heart attack and cancer.

In Kurzweil's second health book, and first with Grossman, the authors explain health issues such as heart disease, cancer, and type 2 diabetes. They promote lifestyle changes such as a low glycemic index diet, calorie restriction, exercise, drinking green tea and alkaline water, using certain supplements, and other changes in daily routines.

Fantastic Voyage states that the purpose of these changes is to obtain and maintain idyllic health with the goal that an individual prolong their life as long as possible. The authors believe that in the coming decades technology will advance to the point of conquering much of the aging process and eliminating degenerative diseases. The book is full of secondary notes on various futuristic topics that show how current research leads to life extension, explaining that future technologies such as bioengineering, nanotechnology and artificial intelligence will change the way we live.

To summarize, the book begins with the description of three "bridges" to the indefinite life. We can simplify and update the information by describing the three bridges as follows, according to our own interpretation:

1. The First Bridge continues into the decade of 2010 and basically consists of doing what your mother or grandmother would tell you (eating well, sleeping well, exercising, not smoking, etc.) with added medical knowledge. This bridge corresponds to Ray and Terry's longevity program (by the first names of Kurzweil and Grossman), including current therapies and guidelines that will allow you to stay healthy long enough to benefit from the construction of the Second Bridge.[18]
2. The Second Bridge will grow strongly during the 2020s with the biotechnology revolution. As we continue to study the genetic code of our biology, we will discover ways to escape disease and aging so that we can fully develop our human potential. This Second Bridge will take us to the Third Bridge.
3. The Third Bridge will correspond mainly to the 2030s and will become a reality thanks to the revolution in nanotechnology and artificial intelligence (AI). The convergence of these technological revolutions will allow us to reconstruct bodies and minds at the molecular level. By 2045, at the latest, we will reach technological singularity and immortality, both biological and computational (i.e. the ability to read, copy and reproduce minds).

Since the sequence of the human genome is making it possible to digitize biology and medicine, *Fantastic Voyage* describes the Second Bridge in the following way[19]:

As we learn how information is transformed in biological processes, many strategies are emerging for overcoming disease and aging processes. We'll review some of the more promising approaches here, and then discuss further examples in the chapters ahead. One powerful approach is to start with biology's information backbone: the genome. With gene technologies, we're now on the verge of being able to control how genes express themselves. Ultimately, we will actually be able to change the genes themselves.

We are already deploying gene technologies in other species. Using a method called recombinant technology, which is being used commercially to provide many new pharmaceutical drugs, the genes of organisms ranging from bacteria to farmyard animals are being modified to produce the proteins we need to combat human diseases.

Another important line of attack is to regrow our cells, tissues, and even whole organs, and introduce them into our bodies without surgery. One major benefit of this therapeutic cloning technique is that we will be able to create these new tissues and organs from versions of our cells that have also been made younger—the emerging field of rejuvenation medicine.

Exponential technologies will contribute to the accelerated development of nanotechnology and AI over the next decade, the first commercial applications of which we will see in the 2030s, leading us to the Third Bridge:

As we "reverse engineer" (understand the principles of operation behind) our biology, we will apply our technology to augment and redesign our bodies and brains to radically extend longevity, enhance our health, and expand our intelligence and experiences. Much of this technological development will be the result of research into nanotechnology, a term originally coined by K. Eric Drexler in the 1970s to describe the study of objects whose smallest features are less than 100 nanometers (billionths of a meter). A nanometer equals roughly the diameter of five carbon atoms.

Robert A. Freitas Jr., a nanotechnology theorist, writes, "The comprehensive knowledge of human molecular structure so painstakingly acquired during the 20th and early 21st centuries will be used in the 21st century to design medically active microscopic machines. These machines, rather than being tasked primarily with voyages of pure discovery, will instead most often be sent on missions of cellular inspection, repair, and reconstruction."

Freitas points out that if "the idea of placing millions of autonomous nanobots (blood cell-sized robots built molecule by molecule) inside one's body might seem odd, even alarming, the fact is that the body already teems with a vast number of mobile nanodevices." Biology itself provides the proof that nanotechnology is feasible. As Rita Colwell, director of the National Science Foundation, has said, "Life is nanotechnology that works." Macrophages (white blood cells) and ribosomes (molecular "machines" that create amino acid strings according to information in RNA strands) are essentially nanobots designed through natural selection. As we engineer our own nanobots to repair and extend biology, we won't be constrained by biology's toolbox. Biology uses a limited set of proteins for all of its creations, whereas we can create structures that are dramatically stronger, faster, and more intricate.

As for the present, Fantastic Voyage offers a series of recommendations to improve our health and reach the Second Bridge alive. In the continuation of that book, *Transcend*, Kurzweil and Grossman propose a more complete program in nine steps according to each letter of the word TRANSCEND[20]:

T: Talk with your doctor.
R: Relaxation.
A: Assessment.
N: Nutrition.
S: Supplements.
C: Calorie Restriction.
E: Exercise.
N: New Technologies.
D: Detoxification.

Mellon and Chalabi's book referred to in the previous chapter (*Juvenescence: Investing in the Age of Longevity*) also includes a series of recommendations that combine the First Bridge with the Second Bridge to keep in mind as we continue to develop the technologies of the Third Bridge towards indefinite longevity, or the new science of "juvenescence", as its authors call it. Juvenescence's recommendations should allow us to reach the escape velocity of longevity in a decade or more, when humanity's greatest industry will emerge, which will be beneficial not only to our health, but also to the global economy and to our personal finances.[21]

Longevity Escape Velocity

The subtitle of Kurzweil and Grossman's *Fantastic Voyage* is very suggestive: *Live Long Enough to Live Forever*. Implicit in this phrase is the idea that if we manage to live long enough in the next few years, until we cross the three bridges and reach rejuvenation, then we could live indefinitely (as long as we want to and that we don't perish due to an accident, a catastrophe, a train at the end of the tunnel, or a piano falling on our heads, among other possible causes of death).

The initial idea is now known as the longevity escape velocity (LEV), which was originally raised by American businessman and philanthropist David Gobel, co-founder of the Methuselah Foundation along with Aubrey de Grey. The concept is based on the planetary escape velocity at which an object, such as a projectile or a rocket, can overcome the force of gravity and leave the Earth. It has been calculated that a speed of 11.2 km/s (kilometres per second) is necessary, which is equivalent to 40,320 km/h (kilometres per hour). Such speed in physics is known as the Earth's planetary escape velocity.[22]

Longevity escape velocity implies a situation in which life expectancy extends more quickly than the time that elapses. For example, when we reach longevity escape velocity, technological advances will annually increase life expectancy by more than a year.

Life expectancy increases slightly each year as treatment strategies and technologies improve. But currently more than 1 year of research is required for each additional year of expected useful life. Longevity escape velocity is reached when this ratio is reversed, so life expectancy increases by more than 1 year for each year of research, as long as that rate of progress is sustainable.

When will it happen? If we look back at history, it is clear that for millennia life expectancy barely increased. It was from the nineteenth century onwards that the great advances in increasing life expectancy began. First days were gained, then weeks, and now months. Today it is estimated that, for every year lived, in the most advanced countries we can increase our life expectancy by 3 months[23]:

> The data shows that in the life expectancy in the leading country of the world has increased by three months every single year.

That is to say, for each year lived, we add 3 more months to our life expectancy. According to Kurzweil, by 2029 we will reach the longevity escape speed, that is, for each year lived, we will gain one more year of life, which means that from that moment we could manage to live indefinitely.[24] As Kurzweil and Grossman say: "live long enough to live forever".

De Grey expresses it with a simple figure, where we can calculate what our life expectancy would be depending on our current age. Unfortunately, for people who are now 100 years old, the outlook is not very promising. For people in their 80 s, the outlook is not good either, unfortunately. But it is likely that people aged 50 or under will reach the escape velocity of longevity, as can be seen in Fig. 4.7.

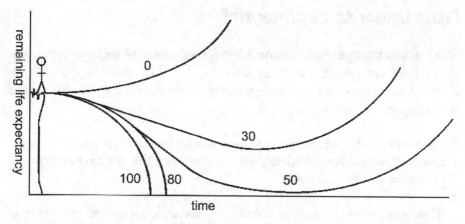

Fig. 4.7 Longevity Escape Velocity (LEV or the "Methuselarity"). (Source: Aubrey de Grey 2008)

There are different opinions on when we will reach the escape velocity of longevity, from very early to never, but 2029 seems reasonable thanks to the exponential advances we are seeing. Indeed, in order to live enough to live forever, we need to pass alive from the Second Bridge to the Third Bridge and increase our healthy life expectancy.[25]

De Grey has also popularized the concept of "Methuselarity" (an original idea of the American entrepreneur Paul Hynek), that is, a kind of Methuselah singularity, which he compares with technological singularity[26]:

> Aging, being a composite of innumerable types of molecular and cellular decay, will be defeated incrementally. I have for some time predicted that this succession of advances will feature a threshold, which I here christen the "*Methuselarity*," following which there will actually be a progressive decline in the rate of improvement in our anti-aging technology that is required to prevent a rise in our risk of death from age-related causes as we become chronologically older. Various commentators have observed the similarity of this prediction to that made by Good, Vinge, Kurzweil and others concerning technology in general (and, in particular, computer technology), which they have termed the "singularity."

Thus, the Methuselarity is a future moment in which all medical conditions that cause human death will be eliminated and death will occur only by accident or homicide. In other words, Methuselarity is the point at which we will reach a period of indefinite life, or without aging, when we reach longevity escape velocity.

From Linear to Exponential

Scientist and businessman Gordon Moore, co-founder of Intel, wrote an article in 1965 explaining that computers doubled their power approximately every 12 months, with far-reaching consequences for computing and related technologies[27]:

> The complexity for minimum component costs has increased at a rate of roughly a factor of two per year… Certainly over the short term this rate can be expected to continue, if not to increase.

This relationship is known as Moore's Law and, as revised by Moore in 1975 with a doubling time of 2 years instead of 1 year, states that approximately every 2 years, the number of transistors in a microprocessor doubles. It is not a law of physics but an empirical observation. It currently applies to personal computers and mobile or cellular phones. However, when it was formulated, there were still no microprocessors (invented in 1971), no personal computers (popularized in the 1980s), nor cellular or mobile telephony (which was barely in the experimental phase).

In his book *The Singularity Is Near: When Humans Transcend Biology*, originally published in 2005, Kurzweil explains that Moore's Law is only part of a much longer historical trend, and with many more expectations for the future.[28] Figure 4.8 shows what Kurzweil calls the Law of Accelerating Returns, where it can be seen that Moore's Law is only one part, corresponding to the fifth paradigm, that reigns at this time. Kurzweil formulated the Law of Accelerating Returns in 2001 and explained then[29]:

> We won't experience 100 years of progress in the twenty-first century — it will be more like 20,000 years of progress (at today's rate).

The Greek philosopher Heraclitus had already said in the fifth century BC that "the only constant thing is change". But today we can see that this change is accelerating, although many do not see it. As Kurzweil explains in his aforementioned book[30]:

> The future is widely misunderstood. Our forebears expected it to be pretty much like their present, which had been pretty much like their past.
>
> [T]he future will be far more surprising than most people realize, because few observers have truly internalized the implications of the fact that the rate of change itself is accelerating.

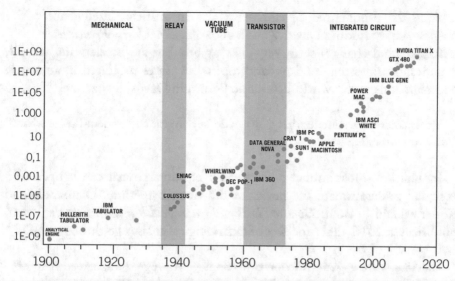

Fig. 4.8 The Law of Accelerating Returns (Calculations per second per $1000). (Source: Ray Kurzweil 2020)

Emphasizing exponential change, Kurzweil explains that there is positive feedback in the technology that accelerates the speed of change[31]:

> Technology goes beyond mere tool making; it is a process of creating ever more powerful technology using the tools from the previous round of innovation.
> [Technological] evolution applies positive feedback…
> The more capable methods resulting from one stage of evolutionary progress are used to create the next stage.
> The first computers were designed on paper and assembled by hand. Today, they are designed on computer workstations with the computers themselves working out many details of the next generation's design, and are then produced in fully automated factories with only limited human intervention.

As technology advances, Kurzweil predicts that in 2029 an artificial intelligence will pass Alan Turing's Test (that is, it will be impossible to distinguish whether one is communicating with a person or with an artificial intelligence) and that in 2045 we will reach the "technological singularity" (which Kurzweil defines simply: the moment when artificial intelligence equals all human intelligence). The Turing Test and technological singularity are not the subjects of this book but, for interested readers, Kurzweil's book *How to Create a Mind* (with a prologue by José Cordeiro in the Spanish edition of Lola Books) explains the exponential developments also in artificial intelligence.[32] You may also enjoy reading David Wood's book *The Singularity Principles*.[33]

Exponential changes seem very slow at first but are accelerating rapidly, American businessman and philanthropist Bill Gates indicated in an article on predictions at the World Economic Forum in Davos, Switzerland[34]:

> Most people overestimate what they can do in one year and underestimate what they can do in ten years.

In addition, most annual predictions overestimate what can happen in a year and underestimate the power of the trend over time. Diamandis and Kotler explain in Bold: *How to Go Big, Create Wealth and Impact the World*, published in 2016, that processes of technological change go through six Ds[35]:

> The Six Ds are a chain reaction of technological progression, a road map of rapid development that always leads to enormous upheaval and opportunity.
>
> Technology is disrupting traditional industrial processes, and they're never going back.
>
> Six Ds are: digitization, deception, disruption, demonetization, dematerialization, and democratization.

According to Diamandis and Kotler, all technologies that can be digitized will undergo exponential transformations that will radically change their respective industries, including medicine and biology, now in the midst of digitization. The six Ds of exponential change begin slowly with digitization and deception and culminate in accelerated dematerialization and democratization of a technology available to all. A classic example is computers, very expensive and slow at the beginning, and now very fast and cheap. The same has happened with mobile phones, which have been democratized globally, so that today everyone, everywhere in the world, has a mobile phone if they wish.

An example applied to biology and medicine is the sequence of the human genome, which began in 1990 with thousands of scientists working in 15 countries around the world. In 1997 only 1% of the human genome had been sequenced, as Kurzweil explains[36]:

> When the human genome scan got underway in 1990 critics pointed out that given the speed with which the genome could then be scanned, it would take thousands of years to finish the project. Yet the fifteen-year project was completed slightly ahead of schedule, with a first draft in 2003.

Year	Cost (US$)	Time
2003	3,000,000,000	13 years
2007	100,000,000	4 years
2008	1,000,000	2 months
2012	10,000	4 weeks
2018	1,000	5 days
2024	100	1 hour
2029	10	1 minute

Fig. 4.9 Time and cost to sequence the human genome. (Source: Authors' own estimates, based on press data and other forecasts 2023)

The reason is very simple. In 1997 only 1% of the total had been sequenced, but the results doubled every year, which meant that in 7 more years, i.e. with 7 more duplications, the sequence of 100% of the genome would be reached, as actually happened. The human genome sequence is an impressive example of exponential technology, both in time and cost. Figure 4.9 shows that if the first human genome sequence cost about $3 billion and took 13 years, the second human genome was completely sequenced 4 years later, in 2007, at an estimated cost of $100 million.

In 2015, a large-scale cost of about a thousand dollars and a week per genome was reached for the first time, and we estimate that in less than a decade the entire genome can be sequenced for as little as 10 dollars in a minute. According to the terminology of Diamandis and Kotler, this will make it possible to move from the first of the six Ds, digitisation, to the last D, democratisation. By the end of the 2020s, everyone, everywhere in the world, will be able to sequence their entire genome and know their predisposition to certain genetic diseases and how to prevent them. In addition, cancer genomes can be sequenced to identify the causes of mutations and attack them directly. We will leave behind procedures such as chemotherapy and radiotherapy to directly locate and eliminate cancerous tumors with high-precision medicine. Chemotherapy and radiotherapy, which are now supposedly modern medicine, will soon become primitive medicine.

Artificial Intelligence Comes to Help

Artificial intelligence will be one of the main technologies that will contribute to understanding biology and improving medicine, also in an exponential way. Artificial intelligence systems already beat humans in chess games (since

1997), in television contests such as Jeopardy (since 2011), in Chinese, Korean and Japanese Go games (since 2016), in Poker games (since 2017), and in reading compression tests (since 2018).[37]

IBM has been one of the historical pioneers in the development of these forms of artificial intelligence, first through its Deep Blue program, which defeated world chess champion Garry Kasparov in 1997, and later with Watson, which beat the Jeopardy champions in front of the television cameras in 2011. IBM later promoted Watson in medical applications, sometimes with the name Doctor Watson, and is approaching human levels in cancer detection and radiological analysis, for example. According to IBM[38]:

> Our purpose is to empower leaders, advocates, and influencers in health through support that helps them achieve remarkable outcomes, accelerate discovery, make essential connections, and gain confidence on their path to solving the world's biggest health challenges.

Google before and Alphabet now are convinced of the power of artificial intelligence to improve the human condition, including health as a priority area. Artificial intelligence developed by the affiliate DeepMind, such as AlphaGo and AlphaZero to play Go, will soon have clinical application. The power of artificial intelligence is truly amazing, as current Google CEO Sundar Pichai explained at a 2018 conference[39]:

> Artificial intelligence is one of the most important things that humanity is working on. It's more profound than, I don't know, electricity or fire.

In that presentation, Pichai did not make direct reference to two other companies under the Alphabet umbrella of Google's companies: Calico (California Life Company) and Verily (the former Google X Life Sciences). Both companies are working in the health field and are increasingly applying Google's "deep learning" technology and other methods to accelerate their business goals. American geneticist and Verily president Andrew Conrad described the difference between Calico and Verily in an interview by science journalist Steven Levy in 2014, when Verily was still Google X Life Sciences[40]:

> Conrad: The mission of Google X Life Sciences is to change healthcare from reactive to proactive. Ultimately it's to prevent disease and extend the average lifespan through the prevention of disease, make people live longer, healthier lives.
> Levy: It sounds like that mission overlaps a little with another Google health enterprise, Calico. Are you working with them?

Conrad: Let me give you the subtle difference. Calico's mission is to improve the maximum lifespan, to make people live longer through developing new ways to prevent aging. Our mission is to make most people live longer, getting rid of the diseases that kill you earlier.

Levy: Basically you're helping me live long enough for Calico's stuff to kick in.

Conrad: Exactly. We're helping you live long enough so Calico can make you live longer.

By 2020, an AI network called AlphaFold developed by DeepMind was able to solve one of biology's most complex problems: protein folding. This was such a landmark, that the prestigious scientific journal *Nature* wrote "it will change everything", and chose AlphaFold as one of the top scientific advances of year.[41] To know how proteins fold will be fundamental to understand how biology works, and AI is thus accelerating our understanding of biology.

According to the terminology of Kurzweil and Grossman, we could simplify by saying that Verily is on the Second Bridge and Calico on the Third Bridge to indefinite lifespans. Besides IBM and Google, other technology companies such as Amazon, Apple, Facebook, GE, Intel and Microsoft, for example, are also developing artificial intelligence that will soon have clinical applications. So are other technology companies in Japan and China, where they are already suffering from problems associated with an aging population. In Japan, large companies such as Sony and Toyota work with robots as health assistants and nurses, and in China, companies such as Baidu (known as China's Google) and BGI (known as *Beijing Genomics Institute* until 2008, and based in the technological city of Shenzhen, near Hong Kong) are developing artificial intelligence focused on disease detection and genomic sequencing.[42]

The Chinese Government has made the strategic decision to become a technological power, from artificial intelligence to medicine and biotechnology. Considering their recent successes, it is likely that they will succeed, and soon, as the national crisis of aging and demographic contraction forces them to do so. China faces the additional problem that its population is beginning to age without having previously enriched itself in relative economic terms. If the advanced countries first became rich and then aged, the opposite is happening in China: the population began to age before it became rich. As if that were not enough, demographic projections indicate that China will suffer an enormous contraction of the population as a result of the "one child" policy. The same is true of Japan, where the population will decline sharply in the coming decades, even though there was never a policy of birth restriction in Japan. That is why it is important for the rest of the world to learn from the experience of Japan and China in relation to their respective demographic crises. Fortunately, the future of the population is not written in stone and current trends can be reversed thanks to the anti-aging and rejuvenation technologies that will be developed in the coming years, many of them precisely in China and Japan.

Artificial intelligence continues to advance rapidly, both in Western and Eastern countries, and one of the first intended uses is for health, medicine, and biology. According to a report by market research firm CB Insights, the fastest growing area for the application of artificial intelligence is health, which is targeted by the largest flow of investments and venture capital. The use of large amounts of data or "macrodata" (*Big Data*), thanks to the proliferation of new personal sensors, many of them medical, will allow more and more information to be analyzed, better comparisons to be made, and the quality of medical diagnoses to be improved. Large companies and small startups are entering the world of medicine thanks to artificial intelligence, including deep learning techniques. Figure 4.10 shows the nascent ecosystem of new medical ventures related to artificial intelligence.

According to the CB Insights report, we will see an accelerated growth of the sector and a significant improvement in people's health thanks to startups that work with exponential technologies that are generating a great disruption in the traditional medical sector[43]:

> We identified over 100 companies that are applying machine learning algorithms and predictive analytics to reduce drug discovery times, provide virtual assistance to patients, and diagnose ailments by processing medical images, among other things.
>
> By 2025, AI systems could be involved in everything from population health management, to digital avatars capable of answering specific patient queries.

Indian-American engineer and businessman Vinod Khosla, co-founder of Sun Microsystems and a venture capitalist in new technologies, explained the exponential changes coming at a Stanford University School of Medicine conference[44]:

> The pace of innovation in software, across all industries, has consistently been much faster than anything else. Within traditional health-care innovation (which intersects with "biological sciences") such as the pharma industry, there are a lot of good reasons those cycles of innovation are slow.
>
> It takes 10 to 15 years to develop a drug and actually be in the marketplace, with an incredibly high failure rate. Safety is one big issue, so I don't blame the

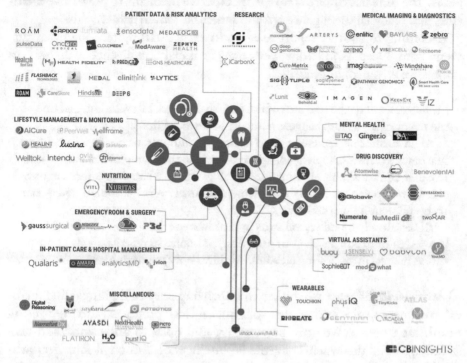

Fig. 4.10 The nascent ecosystem of artificial intelligence for health. (Source: CB Insights 2020)

process. I think it's warranted and the [Food and Drug Administration] is appropriately cautious. But because digital health often has fewer safety effects, and iterations can happen in 2- to 3-year cycles, the rate of innovation goes up substantially.

Over the next ten years data science and software will do more for medicine than all of the biological sciences together.

Several governments have announced that they will begin using new possibilities to improve health through artificial intelligence, new sensors, Big Data and other new technologies. This is the case of the British government, which announced that it will sequencing for free the genome of 500,000 citizens through the UK Biobank from 2020 onward thanks to the support of different companies.[45] The US government has announced a similar measure through the National Institutes of Health to sequence 1000,000 genomes and start its precision medicine initiative in 2022.[46] The Government of Iceland was the first to take such initiatives through the company deCODE in 1996, and later other countries such as Estonia and Qatar have implemented similar plans. The time has come to move from curative medicine to preventive medicine, and artificial intelligence is a fundamental tool to achieve this.

According to another report released in early 2018 by technology investment firm Deep Knowledge Ventures, artificial intelligence will lead to impressive advances in health[47]:

> Healthcare will be the lead area of the Fourth Industrial Revolution, and one of the major catalysts for change is going to be artificial intelligence (AI).
>
> AI in healthcare represents a collection of multiple technologies enabling machines to sense, comprehend, act, and learn so they can perform administrative and clinical healthcare functions. Unlike legacy technologies that are only algorithms/tools that merely complement human work, health AI today can truly augment human activity.
>
> AI has already found several areas in healthcare to revolutionize starting from the design of treatment plans to providing assistance in repetitive jobs to medication management or drug creation. And it is only the beginning.

Artificial intelligence will be key to improving our health, innovating medical treatments, discovering new pharmaceutical products, and optimising healthcare systems. We must be attentive and open to understand and take advantage of all the benefits of artificial intelligence. Although some fear artificial intelligence, we should not see it as a danger, but rather as a great opportunity. Artificial intelligence will complement and increase human intelligence, not replace it. The underlying problem is not artificial intelligence, the real problem is human stupidity, and unfortunately humans are quite stupid by birth, naturally. We believe that with the help of artificial intelligence we will deepen and improve human intelligence and overcome the historical challenge of aging.

From Life Extension to Life Expansion

In Greek mythology, Tithonus was a mortal son of Laomedon, king of Troy, and brother of Priam. Tithonus was so dazzlingly beautiful that the goddess Eos fell in love with him. Eos was the goddess of dawn and asked Zeus to grant immortality to his beloved Tithonus, thus granted by the father of gods. However, the goddess Eos forgot to also ask for eternal youth, so that Tithonus became older and older, shrunken and wrinkled. In some versions, Titono ends up as a cicada or cricket, immortally shrunk and wrinkled.[48]

In this book we defend the extension of life so that we can be indefinitely young, not indefinitely old. The idea is not to survive shrunken and wrinkled like Tithonus, but to live a life of maximum fullness. Making this very clear, it is necessary to move from the extension of life to the expansion of life.

In the introductory section of this book, we mention the Israeli historian Yuval Noah Harari. In his second book, *Homo Deus: A Brief History of Tomorrow*, Harari speaks of immortality as the first great world project of the twenty-first century, and then explains that[49]:

The second big project on the human agenda will probably be to find the key to happiness. Throughout history numerous thinkers, prophets and ordinary people defined happiness rather than life itself as the supreme good. In ancient Greece the philosopher Epicurus explained that worshipping gods is a waste of time, that there is no existence after death, and that happiness is the sole purpose of life. Most people in ancient times rejected Epicureanism, but today it has become the default view. Scepticism about the afterlife drives humankind to seek not only immortality, but also earthly happiness. For who would like to live forever in eternal misery?

For Epicurus the pursuit of happiness was a personal quest. Modern thinkers, in contrast, tend to see it as a collective project. Without government planning, economic resources and scientific research, individuals will not get far in their quest for happiness. If your country is torn apart by war, if the economy is in crisis and if health care is non-existent, you are likely to be miserable. At the end of the eighteenth century the British philosopher Jeremy Bentham declared that the supreme good is 'the greatest happiness of the greatest number' and concluded that the sole worthy aim of the state, the market and the scientific community is to increase global happiness. Politicians should make peace, business people should foster prosperity and scholars should study nature, not for the greater glory of king, country, or God – but so that you and I could enjoy a happier life.

Our goal must be to increase both the quantity of life and the quality of life. It is something that has been happening throughout history. Thousands of years ago life expectancy was between 20 and 25 years. Of that total time, we spent one-third sleeping (assuming 8 hours of sleep in a 24-hour day), and the rest working mainly to survive. In prehistoric times there was no formal education (you learned by working with the older ones, who were not that old, and the learning focused on subsistence work) and there was not much free time left either. This situation remained more or less unchanged for millennia. Even in classical Rome, life expectancy remained in the order of 25 years, as shown in Fig. 4.11.

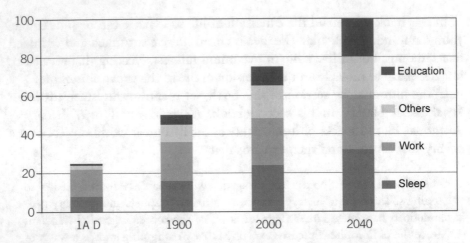

Fig. 4.11 Changing life expectancy throughout history (years by activity). (Source: Authors 2022)

It took several centuries for life expectancy to increase from a quarter of a century (25 years) in the past to about half a century (50 years) at the beginning of the twentieth century. At the beginning of the twenty-first century we have reached an average life expectancy of about three quarters of a century (75 years) and, at the rate we are moving, in a few years we will reach an entire century of life expectancy, which will then increase indefinitely until we reach the longevity escape velocity.

Through all these great transformations of recent centuries, not only has life expectancy increased, but the time available for education and other activities has also increased, far beyond mere subsistence work. It is also important to note that free time has gradually increased throughout history. Thousands of years ago, if we didn't look for food, we died of starvation. If we didn't protect ourselves from animals, we could end up as food for other species. There were no Saturdays or Sundays.

After the invention of agriculture and the founding of the first cities, humans went from nomadic to sedentary, and many religions dedicated a holy day to their god. Thus was born a special day to dedicate to the local deity: some cultures used Saturday, Sunday, or other days. Centuries went by until in the midst of the Industrial Revolution a weekend of two non-working days was created (usually Saturday and Sunday in the European tradition). Now, in the twenty-first century, the first approaches appear to reduce the working day to just four days and/or reduce the hours worked to 30 or 35 hours per week. Thousands of years ago, for our ancestors in Africa this would have been completely unimaginable.

We have come a long way in the extension of life throughout history, and also in the expansion of life. In the last few centuries, our life expectancy has increased greatly, as has the time we can devote to other creative activities. Today we have more time for art, music, sculpture, and many other artistic manifestations than our ancestors ever enjoyed. According to the theories of American psychologist Abraham Maslow, we have climbed the pyramid of human needs. We are leaving behind purely physiological needs in order to concentrate more and more on self-realization needs. This dynamic must continue in the future with more quantity and also more quality of life.[50]

The French philosopher Marie-Jean-Antoine Nicolas de Caritat, known as the Marquis de Condorcet, was a great visionary who lived his last days during the tumultuous period of the French Revolution. His book *Outlines of an Historical View of the Progress of the Human Mind* is an impressive glimpse into the future in a world full of possibilities to come[51]:

Would it be absurd now to suppose that this improvement is capable of indefinite progress; to suppose that the time must come when death will be due only to extraordinary accidents or to the decay (slower and slower down through the generations) of the person's vital forces, and that eventually the amount of time between a person's birth and this decay will have no assignable value?

Humans evolved from other prehuman ancestors millions of years ago, which in turn evolved from earlier, more archaic ancestors, and so on to humble bacteria billions of years ago. So, what is the future of humanity? Now that we are moving from slow biological evolution to rapid technological evolution, will we become quasi-deities as Harari suggests? The English writer William Shakespeare expresses it well in his famous work *Hamlet*[52]:

We know what we are, but know not what we may be.

Human beings have the potential not only to "be" but also to "become". Human beings can use rational means to improve the human condition and the outside world, and we can also use them to improve ourselves, starting with our own body. All these technological opportunities must be put at the service of people to live longer and in better health, to improve our intellectual, physical, and emotional capacities. To paraphrase the popular saying: we are going to add more years to life and also more life to years.

As history shows, we humans have always wanted to transcend our bodily and mental limitations. The way in which these technologies are used will profoundly change the character of our society, and irrevocably alter the vision of ourselves and our place in the grand scheme of things, in the universe, in the very evolution of life. We are beginning a long road to a future full of great opportunities and risks. We must move forward intelligently but without fear, as American science fiction writer David Zindell puts it in his novel *The Broken God*:

— What is a human being, then?'
— 'A seed.'
— 'A ... seed?'
— 'An acorn that is unafraid to destroy itself in growing into a tree.

Notes

1. https://www.amazon.com/Principle-Population-Oxford-Worlds-Classics/dp/0199540454
2. https://www.amazon.com/Leviathan-Oxford-Worlds-Classics-Paperback/dp/B00IIASMRC
3. http://www.diamandis.com/
4. https://www.amazon.com/dp/145161683X
5. http://www.worldbank.org/en/news/feature/2014/04/10/prosperity-for-all-ending-extreme-poverty
6. https://sdgs.un.org/2030agenda
7. https://andrewmcafee.org/more-from-less/overivew
8. https://www.amazon.com/Better-Angels-Our-Nature-Violence/dp/0143122010
9. https://www.amazon.com/Enlightenment-Now-Science-Humanism-Progress/dp/0525427570/
10. http://bigthink.com/in-their-own-words/why-we-love-bad-news-understanding-negativity-bias
11. https://www.amazon.com/Sapiens-Humankind-Yuval-Noah-Harari-ebook/dp/B00ICN066A
12. https://www.amazon.com/population-bomb-Paul-R-Ehrlich-ebook/dp/B071RXJ697
13. https://www.reuters.com/world/china/chinas-population-shrinks-first-time-since-1961-2023-01-17/
14. https://esa.un.org/unpd/wpp/Download/Standard/Population/
15. https://ideas.ted.com/the-genius-of-frugal-innovation/
16. https://ourworldindata.org/future-population-growth
17. https://longevity.technology/news/superagers-initiative-discovering-the-science-behind-exceptional-longevity/

18. https://transcend.me/
19. https://www.amazon.com/Fantastic-Voyage-Live-Enough-Forever/dp/0452286670/
20. https://www.amazon.com/Fantastic-Voyage-Live-Enough-Forever/dp/0452286670/
21. https://www.juvenescence-book.com/book-overview/
22. http://www.mathscareers.org.uk/article/escape-velocities/
23. https://ourworldindata.org/life-expectancy
24. https://singularityhub.com/2017/11/10/3-dangerous-ideas-from-ray-kurzweil/
25. http://www.singularity2050.com/2008/03/actuarial-escap.html
26. https://www.fightaging.org/archives/2009/09/the-methuselarity/
27. https://en.wikiquote.org/wiki/Gordon_Moore
28. https://www.amazon.com/Singularity-Near-Humans-Transcend-Biology-ebook/dp/B000QCSA7C
29. http://www.kurzweilai.net/the-law-of-accelerating-returns
30. https://singularityhub.com/2016/04/05/how-to-think-exponentially-and-better-predict-the-future/
31. https://singularityhub.com/2016/03/22/technology-feels-like-its-accelerating-because-it-actually-is
32. https://www.amazon.com/How-Create-Mind-Thought-Revealed/dp/1491518839
33. https://transpolitica.org/projects/the-singularity-principles/
34. https://www.weforum.org/agenda/2018/01/18-technology-predictions-for-2018/
35. https://www.amazon.com/Bold-Create-Wealth-Impact-World/dp/1476709580
36. https://singularityhub.com/2016/04/05/how-to-think-exponentially-and-better-predict-the-future/
37. https://www.cnet.com/news/new-results-show-ai-is-as-good-as-reading-comprehension-as-we-are/
38. https://www.ibm.com/watson/in-en/about/
39. https://www.theverge.com/2018/1/19/16911354/google-ceo-sundar-pichai-ai-artificial-intelligence-fire-electricity-jobs-cancer
40. https://medium.com/backchannel/were-hoping-to-build-the-tricorder-12e1822e5e6a#
41. https://www.nature.com/articles/d41586-020-03348-4
42. https://www.technologyreview.com/s/609038/chinas-ai-awakening/
43. https://www.cbinsights.com/research/artificial-intelligence-startups-healthcare/
44. http://scopeblog.stanford.edu/2015/05/11/vinod-khosla-shares-thoughts-on-disrupting-health-care-with-data-science/

45. https://www.technologyreview.com/s/609897/500000-britons-genomes-will-be-public-by-2020-transforming-drug-research/
46. https://allofus.nih.gov/
47. http://dkv.global/
48. http://www.perseus.tufts.edu/hopper/text?doc=Apollod.+3.14.3
49. https://www.amazon.com/Homo-Deus-Brief-History-Tomorrow/dp/1910701874
50. https://www.amazon.com/Theory-Human-Motivation-Abraham-Maslow/dp/1614274371
51. https://www.amazon.com/Outlines-Historical-View-Progress-Human/dp/0578016664/
52. https://www.amazon.com/Hamlet-Annotated-Introduction-Charles-Herford/dp/1420952145

5

How Much Does It Cost?

> He who does not value life does not deserve it.
> **Leonardo Da Vinci, 1518**
>
> All my possessions for one moment of time.
> **Queen Elizabeth I Of England, 1603**
>
> Technologies start out affordable only by the wealthy, but at this stage, they actually don't work very well. At the next stage, they're merely expensive, and work a bit better. Then they work quite well and are inexpensive. Ultimately, they're almost free.
> **Ray Kurzweil, 2005**

One concern of many people about the possibility of increasing life expectancy is that this extended longevity will lead to an additional increase in expenditures, especially those related to the weaknesses and diseases of old age. This is a concern that needs to be taken very seriously, especially now that societies are aging.

From Japan to the United States: Rapidly Aging Populations

This concern has been expressed directly on several occasions by the distinguished Japanese politician Taro Aso, grandson of a former Japanese prime minister and himself prime minister from September 2008 to September 2009. During his tenure as prime minister, Aso was the first politician from another continent to visit U.S. President Barack Obama at the White House. Also, during that period, in statements that reverberated around the globe,

© The Author(s), under exclusive license to Springer Nature Switzerland AG 2023
J. Cordeiro, D. Wood, *The Death of Death*, Copernicus Books,
https://doi.org/10.1007/978-3-031-28927-9_5

Aso complained about the cost of taxes to pay for the medical care of pensioners, a social group that needs frequent medical attention[1]:

"I see people aged 67 or 68 at class reunions who dodder around and are constantly going to the doctor," he said. Why should I have to pay for people who just eat and drink and make no effort?

Aso said, not without reason, that these people, who were the same age as him, ought to take more care of themselves, for example walking around every day, and not be so dependent on state aid. In December 2012, after a period in which his party was left out of government and he resigned from the leadership of the party, Aso held two positions: Deputy Prime Minister and Minister of Finance. A month later he would return to the issue of the costs of an aging population in a statement reported in *The Guardian*[2]:

Taro Aso, the finance minister, said on Monday that the elderly should be allowed to "hurry up and die" to relieve pressure on the state to pay for their medical care.

"Heaven forbid if you are forced to live on when you want to die. I would wake up feeling increasingly bad knowing that [treatment] was all being paid for by the government," he said during a meeting of the national council on social security reforms. "The problem won't be solved unless you let them hurry up and die."

Rising welfare costs, particularly for the elderly, were behind a decision last year to double consumption [sales] tax to 10% over the next three years, a move Aso's Liberal Democratic party supported.

To compound the insult, he referred to elderly patients who are no longer able to feed themselves as "tube people".

The journalist also reported on Aso's plans in the event that he himself suffered an illness:

The 72-year-old, who doubles as deputy prime minister, said he would refuse end-of-life care. "I don't need that kind of care," he said in comments quoted by local media, adding that he had written a note instructing his family to deny him life-prolonging medical treatment.

On both occasions, in 2009 and 2012, political correctness forced Aso to quickly rectify his public statements. His advisors were concerned about the possibility of losing the support of the large group made up of the elderly Japanese electorate, the one with the greatest growth and political weight. To say that pensioners were "doddering around" was too blunt, so Aso apologized. He insisted that he did not want to hurt anyone's feelings. On the contrary, he was trying to draw attention to the surging medical costs of illness caused by unhealthy lifestyles. While one must respect one's chosen lifestyle, the publicly funded medical costs of such a choice cannot be allowed to rise indefinitely. That is fair to say.

Aso's comments in Japan were reminiscent of comments made a few decades earlier (1984) by Governor Richard Lamm of Colorado, during a public meeting in Denver. The opinions expressed by Lamm were picked up by *The New York Times*[3]:

Elderly people who are terminally ill have a "duty to die and get out of the way" instead of trying to prolong their lives by artificial means, Gov. Richard D. Lamm of Colorado said Tuesday.

People who die without having life artificially extended are similar to "leaves falling off a tree and forming humus for the other plants to grow up," the Governor told a meeting of the *Colorado Health Lawyers Association* at St. Joseph's Hospital.

"You've got a duty to die and get out of the way," said the 48-year-old Governor.

Lamm's concern was basically the same as Aso's:

The costs of treatment that allows some terminally ill people to live longer are ruining the nation's economic health.

As a society, we make collective decisions to impose limits on personal freedoms. For instance, we insist that everyone wear their seat belts in the car, partly because we want to reduce the medical costs of road traffic injuries. But what about the medical costs resulting from increased life expectancy? Do we really have the right to go on living longer and longer if the costs keep rising as a result?

The Hope that People Will Die Soon

The argument that it would be best for the elderly to "hurry up and die" has been defended not only by several politicians, but also (albeit in more subtle words) by the eminent American medical writer Ezekiel Emanuel. In October 2014, Emanuel wrote an article in *The Atlantic* (254) subtitled "An argument that society and families—and you—will be better off if nature takes its course swiftly and promptly". The title of the article was even more surprising than the subtitle. Emanuel, who was born in 1957, chose the phrase "Why I Hope to Die at 75." In other words, Emanuel hopes to die, without fuss, around the year 2032[4]:

> That's how long I want to live: 75 years.
>
> This preference drives my daughters crazy. It drives my brothers crazy. My loving friends think I am crazy. They think that I can't mean what I say; that I haven't thought clearly about this, because there is so much in the world to see and do. To convince me of my errors, they enumerate the myriad people I know who are over 75 and doing quite well. They are certain that as I get closer to 75, I will push the desired age back to 80, then 85, maybe even 90.
>
> I am sure of my position. Doubtless, death is a loss. It deprives us of experiences and milestones, of time spent with our spouse and children. In short, it deprives us of all the things we value.
>
> But here is a simple truth that many of us seem to resist: living too long is also a loss. It renders many of us, if not disabled, then faltering and declining, a state that may not be worse than death but is nonetheless deprived. It robs us of our creativity and ability to contribute to work, society, the world. It transforms how people experience us, relate to us, and, most important, remember us. We are no longer remembered as vibrant and engaged but as feeble, ineffectual, even pathetic.

Emanuel's credentials are impressive. He has been director of the Department of Clinical Bioethics at the U.S. National Institutes of Health, chairman of the Department of Medical Ethics and Health Policy at the University of Pennsylvania, vice provost of the University of Pennsylvania, and author of the famous book *Reinventing American Health Care: How the Affordable Care Act Will Improve Our Terribly Complex, Blatantly Unjust, Outrageously Expensive, Grossly Inefficient, Error Prone System*, a book that was a staunch defense of President Barack Obama's health initiatives.[5]

Emanuel clearly has great knowledge. As such, he is an important advocate of the aging acceptance paradigm. His point of view deserves our attention. He builds his argument by referring to the case of his father, Benjamin Emanuel, who was also a doctor:

> My father illustrates the situation well. About a decade ago, just shy of his 77th birthday, he began having pain in his abdomen. Like every good doctor, he kept denying that it was anything important. But after three weeks with no improvement, he was persuaded to see his physician. He had in fact had a heart attack, which led to a cardiac catheterization and ultimately a bypass. Since then, he has not been the same.

Once the prototype of a hyperactive Emanuel, suddenly his walking, his talking, his humor got slower. Today he can swim, read the newspaper, needle his kids on the phone, and still live with my mother in their own house. But every thing seems sluggish. Although he didn't die from the heart attack, no one would say he is living a vibrant life. When he discussed it with me, my father said, "I have slowed down tremendously. That is a fact. I no longer make rounds at the hospital or teach."

Here is Emanuel's conclusion:

Over the past 50 years, health care hasn't slowed the aging process so much as it has slowed the dying process. And, as my father demonstrates, the contemporary dying process has been elongated.

The point is that longer life expectancy have led to extended periods of ill health at the end of life. Emanuel refers to quantitative data that support his point of view:

But over recent decades, increases in longevity seem to have been accompanied by increases in disability—not decreases. For instance, using data from the National Health Interview Survey, Eileen Crimmins, a researcher at the University of Southern California, and a colleague assessed physical functioning in adults, analyzing whether people could walk a quarter of a mile; climb 10 stairs; stand or sit for two hours; and stand up, bend, or kneel without using special equipment. The results show that as people age, there is a progressive erosion of physical functioning. More important, Crimmins found that between 1998 and 2006, the loss of functional mobility in the elderly increased. In 1998, about 28 percent of American men 80 and older had a functional limitation; by 2006, that figure was nearly 42 percent. And for women the result was even worse: more than half of women 80 and older had a functional limitation.

The odds of living an unhappy life in old age increase if some stroke statistics are considered:

Take the example of stroke. The good news is that we have made major strides in reducing mortality from strokes. Between 2000 and 2010, the number of deaths from stroke declined by more than 20 percent. The bad news is that many of the roughly 6.8 million Americans who have survived a stroke suffer from paralysis or an inability to speak. And many of the estimated 13 million more Americans who have survived a "silent" stroke suffer from more-subtle brain dysfunction such as aberrations in thought processes, mood regulation, and cognitive functioning. Worse, it is projected that over the next 15 years there will be a 50 percent increase in the number of Americans suffering from stroke-induced disabilities.

In addition, we must consider the challenge of dementia:

The situation becomes of even greater concern when we confront the most dreadful of all possibilities: living with dementia and other acquired mental disabilities. Right now approximately 5 million Americans over 65 have Alzheimer's; one in three Americans 85 and older has Alzheimer's. And the prospect of that changing in the next few decades is not good. Numerous recent trials of drugs that were supposed to stall Alzheimer's—much less reverse or prevent it—have failed so miserably that researchers are rethinking the whole disease paradigm that informed much of the research over the past few decades. Instead of predicting a cure in the foreseeable future, many are warning of a tsunami of dementia—a nearly 300 percent increase in the number of older Americans with dementia by 2050.

The Cost of Aging

Emanuel's view resonates with the opinion expressed in 2003 by Francis Fukuyama, a Japanese American political scientist and professor at Johns Hopkins and Stanford Universities. Fukuyama spoke at a SAGE Crossroads debate entitled "What are the Possibilities and the Pitfalls in Aging Research in the Future?"[6]:

Life extension seems to me a perfect example of something that is a negative externality, meaning that it is individually rational and desirable for any given individual, but it has costs for society that can be negative.

At the age of eighty-five, something like fifty percent of people develop some form of Alzheimer's, and the reason you have this explosion of this particular disease is simply that all of the other cumulative efforts of biomedicine have allowed people to live long enough to where they can get this debilitating disease.

I had a personal experience with this; my mother was in a nursing home for the last couple of years of her life and if you see people caught in that situation it's really a fairly morally troubling thing because nobody wants their loved ones to die, but these people are simply caught in a situation where they have lost control.

American researchers Berhanu Alemayehu and Kenneth E. Warner studied in 2004 what proportions of the expenditure (adjusted for inflation) invested by someone in health services corresponded to each of the stages of his life, results that they published in their report *The Lifetime Distribution of Health Care Costs*.[7] There, they analyzed the health care spending of nearly 4 million members of Blue Cross Blue Shield of Michigan, as well as data from the *Medicare Current Beneficiary Survey*, the *Medical Expenditure Panel Survey*, the *Michigan Mortality Database*, and Michigan nursing home patient counts.

Increased spending on medical services for the elderly can be understood as the result of several factors:

- As people age, they become susceptible to more than one condition at the same time, known as "comorbidity".
- Patients with comorbidity already consume a significant share of national health expenditure due to the complex interactions between different health conditions.
- Even without co-morbidity, an elderly person is less likely to respond quickly to standard medical treatments because their body is weaker and less resilient.
- As the health of the elderly deteriorates, medical science can keep the elderly alive longer than in the past, but in exchange for prolonged treatments, which therefore become more expensive.

This pattern fits into a broader pattern that is sometimes called the "demographic crisis":

- Families have fewer children.
- Senior citizens live longer.
- The proportion of people working continuously decreases, in comparison to those who have left the workforce and who are likely to generate more health costs.
- Without substantial change, national economies face the risk of bankruptcy due to growing demand for health services.

It is in this context that Ezekiel Emanuel proposes his solution: at least some people should voluntarily commit to a date in the future when they will no longer accept expensive healthcare. That could be a date when their lives have already passed through three generations—such as the age of 75 picked for himself by Emanuel.

Emanuel does not advocate euthanasia, assisted suicide, or anything like that, and in fact has a long history of strong opposition to such initiatives. That's not what he has in mind. On the contrary, what he proposes is that:

Once I have lived to 75, my approach to my health care will completely change. I won't actively end my life. But I won't try to prolong it, either. Today, when the doctor recommends a test or treatment, especially one that will extend our lives, it becomes incumbent upon us to give a good reason why we don't want it. The momentum of medicine and family means we will almost invariably get it.

My attitude flips this default on its head. I take guidance from what Sir William Osler wrote in his classic turn-of-the-century medical textbook, The Principles and Practice of Medicine: "Pneumonia may well be called the friend of the aged. Taken off by it in an acute, short, not often painful illness, the old man escapes those 'cold gradations of decay' so distressing to himself and to his friends."

My Osler-inspired philosophy is this: At 75 and beyond, I will need a good reason to even visit the doctor and take any medical test or treatment, no matter how routine and painless. And that good reason is not "It will prolong your life." I will stop getting any regular preventive tests, screenings, or interventions. I will accept only palliative—not curative—treatments if I am suffering pain or other disability.

This means colonoscopies and other cancer-screening tests are out—and before 75. If I were diagnosed with cancer now, at 57, I would probably be treated, unless the prognosis was very poor. But 65 will be my last colonoscopy. No screening for prostate cancer at any age. (When a urologist gave me a PSA test even after I said I wasn't interested and called me with the results, I hung up before he could tell me. He ordered the test for himself, I told him, not for me.) After 75, if I develop cancer, I will refuse treatment. Similarly, no cardiac stress test. No pacemaker and certainly no implantable defibrillator. No heart-valve replacement or bypass surgery. If I develop emphysema or some similar disease that involves frequent exacerbations that would, normally, land me in the hospital, I will accept treatment to ameliorate the discomfort caused by the feeling of suffocation, but will refuse to be hauled off.

What about simple stuff? Flu shots are out. Certainly if there were to be a flu pandemic, a younger person who has yet to live a complete life ought to get the vaccine or any antiviral drugs. A big challenge is antibiotics for pneumonia or skin and urinary infections. Antibiotics are cheap and largely effective in curing infections. It is really hard for us to say no. Indeed, even people who are sure they don't want life-extending treatments find it hard to refuse antibiotics. But, as Osler reminds us, unlike the decays associated with chronic conditions, death from these infections is quick and relatively painless. So, no to antibiotics.

Obviously, a do-not-resuscitate order and a complete advance directive indicating no ventilators, dialysis, surgery, antibiotics, or any other medication—nothing except palliative care even if I am conscious but not mentally competent—have been written and recorded. In short, no life-sustaining interventions. I will die when whatever comes first takes me.

Clash of Paradigms

Emanuel's perspective can be described as courageous and selfless. Moreover, it is coherent with the paradigm he uses to interpret the world:

- Medical costs continue to rise due to the elderly and society is less and less able to bear those costs.
- Long-standing hopes for progress in curing diseases such as dementia have proven unfounded.
- The elderly, suffering from long age-related illnesses, enjoy a poor quality of life.
- Society needs a rational and humane strategy to allocate its limited health resources.
- The elderly have already lived the best years of their lives, they have already spent their moment of maximum productivity and creativity.

Regarding the last point, Emanuel quotes the famous scientist Albert Einstein:

> But the fact is that by 75, creativity, originality, and productivity are pretty much gone for the vast, vast majority of us. Einstein famously said, "A person who has not made his great contribution to science before the age of 30 will never do so."

Note, however, that Emanuel feels compelled to contradict Einstein immediately and then reiterates his own opinion in a less radical form:

> [Einstein] was extreme in his assessment. And wrong. Dean Keith Simonton, at the University of California at Davis, a luminary among researchers on age and creativity, synthesized numerous studies to demonstrate a typical age-creativity curve: creativity rises rapidly as a career commences, peaks about 20 years into the career, at about age 40 or 45, and then enters a slow, age-related decline. There are some, but not huge, variations among disciplines. Currently, the average age at which Nobel Prize–winning physicists make their discovery—not get the prize—is 48. Theoretical chemists and physicists make their major contribution slightly earlier than empirical researchers do. Similarly, poets tend to peak earlier than novelists do. Simonton's own study of classical composers shows that the typical composer writes his first major work at age 26, peaks at about age 40 with both his best work and maximum output, and then declines, writing his last significant musical composition at 52.

Moreover, Emanuel is also forced to cite some counterexamples afterwards:

About a decade ago, I began working with a prominent health economist who was about to turn 80. Our collaboration was incredibly productive. We published numerous papers that influenced the evolving debates around health-care reform. My colleague is brilliant and continues to be a major contributor, and he celebrated his 90th birthday this year. But he is an outlier—a very rare individual.

Emanuel acknowledges these counterexamples to his general opinion, but implies that such counterexamples are rare, on account of the complexity of the brain and the decline in so-called "brain plasticity":

The age-creativity curve—especially the decline—endures across cultures and throughout history, suggesting some deep underlying biological determinism probably related to brain plasticity.

We can only speculate about the biology. The connections between neurons are subject to an intense process of natural selection. The neural connections that are most heavily used are reinforced and retained, while those that are rarely, if ever, used atrophy and disappear over time. Although brain plasticity persists throughout life, we do not get totally rewired. As we age, we forge a very extensive network of connections established through a lifetime of experiences, thoughts, feelings, actions, and memories. We are subject to who we have been. It is difficult, if not impossible, to generate new, creative thoughts, because we don't develop a new set of neural connections that can supersede the existing network. It is much more difficult for older people to learn new languages. All of those mental puzzles are an effort to slow the erosion of the neural connections we have. Once you squeeze the creativity out of the neural networks established over your initial career, they are not likely to develop strong new brain connections to generate innovative ideas—except maybe in those Old Thinkers like my outlier colleague, who happen to be in the minority endowed with superior plasticity.

In response to the question as to why medicine cannot make many more people experience the same kind of increased creativity and productivity that he calls "outliers", Emanuel again relies on one of the points of his paradigm, that is, that old hopes of making progress in curing diseases such as dementia have proved unfounded.

Paradigm Shift

It is not surprising that the different points of that paradigm fit well together and reinforce each other. That is where paradigms draw their strength. However, the goal of reducing health care spending for the elderly can be achieved in a very different way: through ideas that anticipate the paradigm of rejuvenation. If it turns out to be true that intelligent and concentrated medical research can delay the onset and consequences of aging (perhaps indefinitely), society would benefit greatly. In fact, more people will:

- Stop aging and weakening
- No longer be victims of age-related diseases (including diseases such as cancer and cardiovascular disease, the likelihood and severity of which increase with age)
- Stop consuming large amounts of medical services derived from long-term illness
- Continue to be an active and productive part of the workforce and will retain their vigor and enthusiasm.

Therefore, short-term investments will result in substantial financial and social benefits through improved health and delayed aging. This is known as the "longevity dividend".

The "Longevity Dividend"

The concept of the "Longevity Dividend" was introduced in a 2006 article in the scientific journal *The Scientist*, "In Pursuit of the Longevity Dividend". The article was written by a quartet of experienced researchers from various areas of aging: S. Jay Olshansky, professor of epidemiology and biostatistics at the University of Illinois, Chicago; Daniel Perry, then executive director of the Alliance for Aging Research in Washington, DC; Richard A. Miller, professor of pathology at the University of Michigan, Ann Arbor; and Robert N. Butler, president and CEO of the International Longevity Center in New York. The article contains an urgent call[8]:

> We suggest that a concerted effort to slow aging begin immediately—because it will save and extend lives, improve health, and create wealth.

The last of these reasons is worth highlighting: this effort to slow aging *will create wealth*.

The authors of the article are optimistic about the scientific perspectives of anti-aging:

> In recent decades biogerontologists have gained significant insight into the causes of aging. They've revolutionized our understanding of the biology of life and death. They've dispelled long-held misconceptions about aging and its effects, and offered for the first time a real scientific foundation for the feasibility of extending and improving life.
>
> The idea that age-related illnesses are independently influenced by genes and/ or behavioral risk factors has been dispelled by evidence that genetic and dietary interventions can retard nearly all late-life diseases in parallel. Several lines of evidence in models ranging from simple eukaryotes to mammals suggest that our own bodies may well have "switches" that influence how quickly we age. These switches are not set in stone; they are potentially adjustable.
>
> Nevertheless, the belief that aging is an immutable process, programmed by evolution, is now known to be wrong. In recent decades, our knowledge of how, why, and when aging processes take place has progressed so much that many scientists now believe that this line of research, if sufficiently promoted, could benefit people alive today. Indeed, the science of aging has the potential to do what no drug, surgical procedure, or behavior modification can do-extend our years of youthful vigor and simultaneously postpone all the costly, disabling, and lethal conditions expressed at later ages.

As a result, the four researchers anticipate many advantages, including "enormous economic benefits":

> In addition to the obvious health benefits, enormous economic benefits would accrue from the extension of healthy life. By extending the time in the lifespan when higher levels of physical and mental capacity are expressed, people would remain in the labor force longer, personal income and savings would increase, age-entitlement programs would face less pressure from shifting demographics, and there is reason to believe that national economies would flourish. The science of aging has the potential to produce what we refer to as a "Longevity Dividend" in the form of social, economic, and health bonuses both for individuals and entire populations—a dividend that would begin with generations currently alive and continue for all that follow.

The authors go on to list the different ways in which healthy life extension would create wealth, both for individuals and for the societies in which they live:

- Healthy seniors accumulate more savings and investments than those afflicted by disease.
- They tend to stay productive in society.
- They trigger economic booms in so-called mature markets, including financial services, tourism, hospitality, and intergenerational transfers to younger generations.
- Improvements in health status also lead to less school and work absenteeism and are associated with better education and higher incomes.

However, the authors also consider the alternative scenario, in which research on rejuvenation therapies is under-resourced and progresses very slowly. In this scenario, age-related diseases would demand from society an ever-increasing amount of spending:

> Consider what is likely to happen if we don't. Take, for instance, the impact of just one age-related disorder, Alzheimer disease (AD). For no other reason than the inevitable shifting demographics, the number of Americans stricken with AD will rise from 4 million today to as many as 16 million by midcentury. This means that more people in the United States will have AD by 2050 than the entire current population of the Netherlands.
>
> Globally, AD prevalence is expected to rise to 45 million by 2050, with three of every four patients with AD living in a developing nation. The US economic toll is currently $80-$100 billion, but by 2050 more than $1 trillion will be spent annually on AD and related dementias. The impact of this single disease will be catastrophic, and this is just one example.
>
> Cardiovascular disease, diabetes, cancer, and other age-related problems account for billions of dollars siphoned away for "sick care." Imagine the problems in many developing nations where there is little or no formal training in geriatric health care. For instance, in China and India the elderly will outnumber the total current US population by midcentury. The demographic wave is a global phenomenon that appears to be leading health care financing into an abyss.

In other words, these researchers anticipate the same financial crisis exposed by Ezekiel Emanuel. However, while Emanuel recommends a (voluntary) withdrawal of expensive medical help once a certain age is reached, for example at age 75, these four authors believe that the science of anti-aging can provide a better solution that does not involve any cessation of medical help:

> Nations may be tempted to continue attacking diseases and disabilities of old age separately, as if they were unrelated to one another. This is the way most medicine is practiced and medical research is conducted today. The National

Institutes of Health in the United States are organized under the premise that specific diseases and disorders be attacked individually. More than half of the National Institute on Aging budget in the United States is devoted to AD. But the underlying biological changes that predispose everyone to fatal and disabling diseases and disorders are caused by the processes of aging. It therefore stands to reason that an intervention that delays aging should become one of our highest priorities.

Such interventions are, of course, the subject of this book. We defend the prediction that relatively soon we will have treatments that will allow us to extend indefinitely the healthy life expectancy. The proponents of the longevity dividend point out that, even if the extension does not become indefinite—if, for example, it only leads to seven more years of healthy life—it would still be extraordinarily positive from both an economic and a humanitarian perspective:

> We envision a goal that is realistically achievable: a modest deceleration in the rate of aging sufficient to delay all aging-related diseases and disorders by about seven years. This target was chosen because the risk of death and most other negative attributes of aging tends to rise exponentially throughout the adult lifespan with a doubling time of approximately seven years. Such a delay would yield health and longevity benefits greater than what would be achieved with the elimination of cancer or heart disease. And we believe it can be achieved for generations now alive.
>
> If we succeed in slowing aging by seven years, the age-specific risk of death, frailty, and disability will be reduced by approximately half at every age. People who reach the age of 50 in the future would have the health profile and disease risk of today's 43-year-old; those aged 60 would resemble current 53-year-olds, and so on. Equally important, once achieved, this seven-year delay would yield equal health and longevity benefits for all subsequent generations, much the same way children born in most nations today benefit from the discovery and development of immunizations.

Quantifying the Longevity Dividend

Against the idea of the longevity dividend, three main arguments are usually considered, specifically:

1. The first is the absolutist position that no accumulation of research is going to extend healthy longevity in humans by that seven-year period. This

position argues that improvements similar to those of the past cannot be repeated in the present, regardless of the level of investment made.

2. The second argument is that such research will be extremely expensive, so that the potential economic benefits from longer healthy life expectancies will be offset by the enormous costs of obtaining such benefits.

3. Finally, the third argument argues that the benefits of the longevity dividend are only temporary: the significant medical expenditure for the elderly is not cancelled, only postponed.

We reject outright the first argument that in healthy longevity "no more major breakthroughs will ever be achieved". On the contrary, what remains to be seen is "how much", "how fast" and "at what cost". This brings us to the second argument. It is an argument that deserves more attention, so we should try to quantify the figures at stake.

One way to address the numbers is in an article written by American scholars Dana Goldman, David Cutler, and others under the title *"Substantial Health and Economic Returns from Delayed Aging May Warrant a New Focus for Medical Research"*. Goldman is professor of public policy and pharmaceutical economics and director of the Schaeffer Center for Health Policy & Economics at the University of Southern California, while Cutler is professor of economics at Harvard University.[9]

These authors begin by saying that, if health systems continue on their present trajectory, spending on health insurance for the elderly (the Medicare that provides health insurance to Americans aged 65 and over) will rise from 3.7% of U.S. GDP in 2012 to a gigantic 7.3% in 2050. This reflects the greater amount of time older people spend in a state of disability compared to the past:

> Although attacking diseases has extended life for younger and middle-aged people, evidence suggests it may not extend healthy life once people reach older ages. Increased disability rates are now accompanying increases in life expectancy, leaving the length of a healthy life span unchanged or even shorter than in the past.
>
> As people age, they are now much less likely to fall victim to a single isolated disease than was previously the case. Instead, competing causes of death more directly associated with biological aging (for example, heart disease, cancer, stroke, and Alzheimer's disease) cluster within individuals as they reach older ages. These conditions elevate mortality risk and create the frailty and disabilities that can accompany old age.

The authors study four different scenarios. They all arise depending on the different types of medical progress that may occur in the period between 2010 and 2050:

- The "status quo scenario", in which disease mortality rates do not change during the indicated period.
- A "delayed cancer scenario", in which cancer incidence is reduced by 25% between 2010 and 2030, and then remains constant.
- A "delayed heart disease scenario," in which the incidence of heart disease is reduced by 25% between 2010 and 2030, and then kept constant.
- A "delayed aging scenario" in which "mortality from factors such as age, as opposed to exposure to external risks such as trauma or smoking… would decline by 20% by 2050".

The fourth of these scenarios fits the idea that this book defends. As the authors describe it:

Although this scenario altered the effects of getting disease, it was not the same as scenarios of disease prevention because it addressed the underlying biology of aging. The scenario reduced mortality and the probability of onset of both chronic conditions (heart disease, cancer, stroke or transient ischemic attack, diabetes, chronic bronchitis and emphysema, and hypertension) and disability by 1.25 percent for each year of life lived above age fifty (the period in life when most of these diseases emerge). This reduction was phased in over twenty years, starting with a 0 percent reduction in 2010 and increasing linearly until the full 1.25 percent reduction was achieved in 2030.

All three intervention scenarios lead to an increase in life expectancy. Someone who is 51 years old in 2030 would have a life expectancy of 35.8 years (status quo scenario), 36.9 years (delayed cancer scenario), 36.6 years (delayed cardiovascular disease scenario), or 38.0 years (delayed aging scenario). The delayed aging scenario is the one that goes the furthest because it affects all age-related diseases, while in the other two cases people remain vulnerable to all diseases, with the exception of the disease particularly treated in each of the interventions.

Increases in life expectancy are modest, only about 1 year in the specific disease scenarios and 2.2 years in the delayed aging scenario. What is much more striking, however, are the financial consequences of these delays in each of the models studied. Including the expected costs generated by public programmes such as health care for elderly people, health care for disadvantaged people, disability insurance, social security premiums, etc., and including estimates of productivity gains from better living conditions, the authors estimate that the economic value of the aging delay scenario would be $7.1 trillion over the period to 2060. This benefit has two sources:

1. Fewer disabled seniors, up to 5 million fewer in the United States for each of the years between 2030 and 2060.
2. An increase in the number of non-disabled seniors, up to 10 million more in the United States during the period considered, resulting in greater contributions to the economy (both in terms of production and consumption).

As the differences that would be achieved in the other two scenarios (delayed cancer and delayed heart disease) are much smaller, the benefits derived from them are also much smaller. This is another reason why priority should be given to rejuvenation rather than continuing to treat diseases on an individual basis.

It is inevitable that there will be many doubts regarding the figures mentioned. However, even if the leading figure of $7.1 trillion were significantly wrong, the advantages would still be very significant. And what is especially interesting is that these benefits arise from a very small increase in life expectancy of just 2.2 years. Imagine how big the benefits of a larger increase could be.

Financial Benefits from Living Longer Lives

It should be noted that the savings described in the previous section depend on major changes in the rules determining entitlement to welfare state pensions. As Goldman, Cutler and their colleagues point out:

> Delayed aging would greatly increase entitlement outlays, especially for Social Security. However, these changes could be offset by increasing the Medicare eligibility age and the normal retirement age for Social Security.

Without changes in the start and timing of pension payments, more life years would increase existing financial difficulties. The extent of these problems was highlighted in a 2012 report by the International Monetary Fund, as summarised in a Reuters article by Stella Dawson entitled "*Cost of aging rising faster than expected-IMF*"[10]:

People worldwide are living three years longer than expected on average, pushing up the costs of aging by 50 percent, and governments and pension funds are ill prepared, the International Monetary Fund said.

Already the cost of caring for aging baby boomers is beginning to strain government budgets, particularly in advanced economies where by 2050 the elderly will match the numbers of workers almost one for one. The IMF study shows that the problem is global and that longevity is a bigger risk than thought.

If everyone in 2050 lived just three years longer than now expected, in line with the average underestimation of longevity in the past, society would need extra resources equal to 1 to 2 percent of GDP per year.

For private pension plans in the United States alone, an extra three years of life would add 9.0 percent to liabilities, the IMF said in urging governments and the private sector to prepare now for the risk of longer lifespans.

We are therefore talking about huge numbers:

To give an idea of how costly this could prove, the IMF estimated that if advanced economies were to plug the shortfall in pension savings of an extra three years immediately, they would have to stash away the equivalent of 50 percent of 2010 GDP, and emerging economies would need 25 percent.

These extra costs fall on top of the doubling in total expenses that countries can expect through 2050 from an aging population. The faster countries tackle the problem, the easier it will be to handle the risk of people living longer, the IMF said.

However, this report does not speak of two fundamental things to consider in any forward-looking vision:

1. The potential of longer-lived people to contribute more to the economy (and not represent a drain on resources).
2. The possibility of changing the start age of receiving pension benefits in order to harmonise it with changes in average life expectancy.

A similar argument is made by economists Henry Aaron and Gary Burtless of the Brookings Institution in Washington, DC, in their book *Closing the Deficit: How Much Can Later Retirement Help?*[11] Their conclusions are summarized by Walter Hamilton in the *Los Angeles Times*[12]:

The book points out that people older than 60 have been steadily delaying retirement over the last 20 years. From 1991 to 2010, the employment rate increased by more than half among 68-year-old men and by about two-thirds among women the same age.

As people… work longer in life, they'll generate additional tax revenue that could reduce federal budget deficits and spending on Social Security.

The increase in work could boost government revenue by as much as $2.1 trillion over the next three decades.

Expenditures on Social Security and Medicare could decrease by more than $600 billion as people delay tapping into those programs. The total effect, including savings on interest from smaller annual deficits, could narrow the gap between government revenue and spending by more than $4 trillion through 2040.

William Nordhaus, the famous American economist at Yale University, had more or less reached the same conclusion in his 2002 publication "*The Health of Nations: The Contribution of Improved Health to Living Standards*".[13] Nordhaus analyzed the causes of the improvements in economic performance during the twentieth century, and his conclusion was that the increase in life expectancy is, in terms of increased economic performance, "about as large as the value of all other consumption goods and services put together". As they live longer, people work longer, produce more, and provide more experience to the workforce and the community as a whole. At the end of his study, Nordhaus summarizes his theses as follows:

> To a first approximation, the economic value of increases in longevity in the last hundred years is about as large as the value of measured growth in non-health goods and services.

Kevin Murphy and Robert Topel, two prestigious economists at the University of Chicago, make another calculation of historical gains from greater longevity in their 2005 article "The Value of Health and Longevity. These economists make an extensive calculation, in an article that is 60 pages long, but their conclusions can be read in the synopsis[14]:

> We develop an economic framework for valuing improvements to health and life expectancy, based on individuals' willingness to pay. We then apply the framework to past and prospective reductions in mortality risks, both overall and for specific life-threatening diseases. We calculate the social values of increased longevity for men and women over the 20th century; the social value of progress against various diseases after 1970; and the social value of potential future progress against various major categories of disease. The historical gains from increased longevity have been enormous. Over the 20th century, cumulative gains in life expectancy were worth over $1.2 million per person for both men and women. Between 1970 and 2000 increased longevity added about $3.2 trillion per year to national wealth, an uncounted value equal to about half of average annual GDP over the period. Reduced mortality from

heart disease alone has increased the value of life by about $1.5 trillion per year since 1970.

Murphy and Topel look forward to these gains from new health improvements continuing into the future:

> The potential gains from future innovations in health care are also extremely large. Even a modest 1 percent reduction in cancer mortality would be worth nearly $500 billion.

However, two fundamental issues remain outstanding:

1. Will the costs of achieving this healthy longevity extension outweigh the economic benefit of (perhaps) trillions of dollars?
2. Will additional years of healthy longevity flow into particularly expensive years of health care so that problems are only postponed and not solved?

Let us answer these two questions in order.

Rejuvenation Therapy Development Costs

It is not possible to know in advance precisely what it would cost to develop rejuvenation therapies that would extend healthy life expectancy by an average of, say, 7 years (as proposed in the 2006 article mentioned above, "In Pursuit of the Longevity Dividend," by Olshansky and his colleagues). There are too many unknowns involved to even reach any plausible "order of magnitude" estimate. We do not know how difficult it will be to resolve the cellular and molecular keys to age-related diseases. However, we can develop some confidence from observing various past projects to extend the healthy life, as they generally managed to cover their costs with ease. For example, consider programmes to vaccinate children against childhood-related diseases. The basic principle is that prevention can be much cheaper than cure. According to American scientist Brian Kennedy, who was executive director of the *Buck Institute* for Research on Aging in California, "prevention costs can be as much as one-twentieth of treatment costs".[15]

Murphy and Topel, whose research were mentioned in the previous section, make the following overall assessment:

Between 1970 and 2000 increased longevity yielded a "gross" social value of $95 trillion, while the capitalized value of medical expenditures grew by $34 trillion, leaving a net gain of $61 trillion… Overall, rising medical expenditures absorb only 36% of the value of increased longevity.

The authors indicate the implications of their analysis to determine the level of future investment in health innovation:

An analysis of the social value of improvements in health is a first step toward evaluating the social returns to medical research and health-augmenting innovations. Improvements in health and longevity are partially determined by society's stock of medical knowledge, for which basic medical research is a key input. The U.S. invests over $50 billion annually in medical research, of which about 40 percent is federally funded, accounting for 25 percent of government research and development outlays. The $27 billion federal expenditure for health-related research in FY 2003, the vast majority of which is for the National Institutes of Health, represented a real dollar doubling over 1993 outlays. Are these expenditures warranted?

Our analysis suggests that the returns to basic research may be quite large, so that substantially greater expenditures may be worthwhile. By way of example, take our estimate that a 1 percent reduction in cancer mortality would be worth about $500 billion. Then a "war on cancer" that would spend an additional $100 billion (over some period) on cancer research and treatment would be worthwhile if it has a 1-in-5 chance of reducing mortality by 1 percent, and a 4-in-5 chance of doing nothing at all.

Note the mention of probabilities. An investment may make sense even if the probability of success is relatively small. This is something that venture capital managers understand well. They are willing to accept small probabilities of success in a company's business objectives provided the amount of success (if any) is large enough. A 5% probability of an eventual multi-billion capitalization of a company would be a good investment if, for example, its future valuation would exceed its present valuation by 100 times or more.

These types of considerations are familiar to anyone who evaluates insurance policies. It is reasonable to think that the most unlikely disasters should also be covered by insurance policies.

If the small odds deserve our attention because the consequences of them happening are important enough, why are we not going to pay even more attention to the possibility of something happening with a 50% probability and with financial consequences that, if carried out, would reach trillions of dollars? Considering the figures associated with the most satisfactory scenario, that would be the situation if the rejuvenation programme were successful, even on a very small scale.

In the journal *Nature Aging*, which is a new publication by the *Nature* group dedicated exclusively to aging research, Australian biologist David A. Sinclair at Harvard University, together with British economists Andrew J. Scott at the London School of Economics and Martin Ellison at the University of Oxford, published in 2021 a very important article quantifying "The economic Value of Targeting Aging" and estimating annual savings of US$38 trillion by extending life expectancy by just 1 year[16]:

> Developments in life expectancy and the growing emphasis on biological and 'healthy' aging raise a number of important questions for health scientists and economists alike. Is it preferable to make lives healthier by compressing morbidity, or longer by extending life? What are the gains from targeting aging itself compared to efforts to eradicate specific diseases? Here we analyze existing data to evaluate the economic value of increases in life expectancy, improvements in health and treatments that target aging. We show that a compression of morbidity that improves health is more valuable than further increases in life expectancy, and that targeting aging offers potentially larger economic gains than eradicating individual diseases. We show that a slowdown in aging that increases life expectancy by 1 year is worth US$38 trillion, and by 10 years, US$367 trillion. Ultimately, the more progress that is made in improving how we age, the greater the value of further improvements.

Additional Sources of Funding

There are at least five potential sources of funding that would accelerate rejuvenation therapies and thus the achievement of the longevity dividend.

First, let us consider all the funding that is now devoted to fighting individual diseases and compare it with that which is devoted to addressing the underpinning mechanisms of aging. Of the nearly $30 billion annual medical research budget monitored by the U.S. National Institutes of Health, less than 10% is currently spent on aging, with the remainder spread among individual diseases.[17] This current pattern of distribution of funding, which is also repeated in the health budgets of many countries, is in line with the dominant strategy that says that to improve health, "diseases come first". However, if aging were to receive a larger share of the total budget (perhaps as much as 20% over the next 10 years rather than less than 10%), then many diseases could cease to be so prevalent and so severe, despite the reduction in research funds earmarked specifically for them. This would mean taking on an alternative strategy to improve health, one that says "aging comes first ", as aging increases the body's tendency to suffer from diseases and increases the likelihood of complications from these diseases.

A second way to achieve greater progress in rejuvenation is by an increase in the proportion of people's discretionary time that is spent in research into anti-aging therapies. Only a small increase would be needed in the percentages of time of individual researchers so that at an aggregate level there would be a great increase across the entire population. If only one person out of a thousand devoted just four more hours a week to rejuvenation research, and therefore 4 h less to leisure activities such as watching TV entertainment, the total number of hours dedicated to rejuvenation in a country could skyrocket. Much of that effort will be of little importance in absolute terms if it is devoted to reviewing what others have already done and if the people involved have restricted access to experimental material and facilities. However, if suitable frameworks and processes for "collaborative engineering of rejuvenation" are put in place, including educational and orientation activities, the global benefit could be substantial.

Third. The alternative to devoting more of your time is for people around the world to donate more of their savings to rejuvenation research initiatives. For example, instead of donating to the university where they were educated or to their local parish, these funds (or part of them) could be redirected to philanthropic organizations in the anti-aging field. These investments could be seen as a kind of parallel contribution to pension plans and insurance policies: the more people donate, the less likely it is that family members, neighbors and others close to them will suffer from age-related illnesses. If, following this book, there is a drastic change in the way public opinion acts, we could see an increase in this type of funding, in the same way that has occurred with campaigns that have become widespread (such as pink ribbons for the fight against breast cancer).

Fourth. Companies (both large and small) may decide to invest in this field because of the potential financial benefit of participating in the longevity dividend. After all, if these therapies succeed in generating a higher level of wealth in society by increasing productive economic activity and decreasing prolonged absences from work, there should be ways for the companies providing these therapies to receive part of the new wealth generated. If such benefit sharing could be concretized and specified, a substantially greater share of the great entrepreneurial capacity of the business world would help the cause of rejuvenation.

Fifth. It is time to address the issue of increasing public funds rather than simply relocating existing public funds for health. Public funds can often address issues not covered by private company funds. Public investment can afford to be more patient with expected returns. Profits go to the whole of society; they are not redirected to shareholders or executives. One example was the large U.S. contribution with the Marshall Plan ($13 billion during the 1940s), a financial aid program to rebuild Western Europe after the devastation caused by World War II. Two other examples would be the atomic Manhattan Project to win World War II and the Project Apollo during the Cold War to get the first human to the Moon.

Another comparable case would be the British public funding of the National Health Service. Also, the European investment in CERN, the European Organisation for Nuclear Research, which has a large hadron collider (LHC). This multibillion-euro investment over several decades was not undertaken with short-term economic benefits in mind. On the contrary, politicians supported CERN on the basis of an overall vision aimed at gathering fundamental information about the natural world, and perhaps generating economic benefits in the future in ways that are difficult to anticipate. The CERN Higgs boson detection project alone is estimated to have already consumed some $13.25 billion.[18] Even the Internet was born out of Tim Berners-Lee's work at CERN between 1989 and 1991. However, there are good reasons to downgrade the priority of several public initiatives such as CERN (several more examples could be given) over the next few decades and instead increase public research funding for rejuvenation.

In conclusion, there are several possible sources of resources to make a significant additional effort that can be applied to the rejuvenation project in the expectation that at least part of that effort will end up producing huge economic benefits. Society has an important decision to make regarding the priorities of these resources and the scale on which they should be invested.

Curing Aging Will Be Cheaper People Believe

We have already seen that there are multiple ways of financing, both public and private, which depend on the decisions of governments and entrepreneurs, with the fundamental support of the citizens, because aging is a disease that affects all humanity. We must never forget that aging is the leading cause of death in the world.

We have also explained how medicine has, up till now, focused more on attacking the symptoms rather than the causes of aging. We now need a real preventive medicine, and not so much curative, to avoid the processes of aging. Instead of spending $7 billion on curing disease, especially in the painful final stages of people's lives, we need to invest that amount in preventing the aging process early on.

If we analyze the basic fundamentals of human beings, in terms of our basic chemical composition we can say that we are quite simple. An adult human is composed of about 60% water (although much depends on age, sex and adiposity, among other factors). Moreover, we are not even Evian or Perrier water, we are very ordinary water composed of H_2O, i.e. two hydrogen atoms and one oxygen atom. We have organs with more water and others with less water: for example, bones are estimated to contain 22% water, muscles and brain 75%, heart 79%, blood and kidneys 83%, and liver 86%.[19] Water proportions also change greatly with age. Children have up to 75% water, adults 60% and the elderly 50%. According to Nestlé Waters, an average adult of 60 kilograms has 42 liters of water in his body, distributed as follows[20]:

- 28 liters of intracellular water:
- 14 liters are found in the extracellular fluid, of which:
  - 10 liters of interstitial liquid (including lymph), which is an aqueous medium surrounding the cells.
  - 3 liters is the blood plasma.
  - 1 liter is transcellular fluid (cerebrospinal, ocular, pleural, peritoneal and synovial fluid).

Apart from being mainly water, which in addition to oxygen contains the most abundant element in the entire universe, hydrogen, the rest of the human body is composed of few chemical elements, relatively abundant and cheap. Only four basic elements (oxygen, carbon, hydrogen, and nitrogen) represent 99% of the total of all atoms and 96% of the weight of an average human of average age and 70 kg of weight, as can be clearly seen in Table 5.1.[21]

Although we have fewer oxygen than hydrogen atoms, oxygen atoms (with atomic number 8, i.e. oxygen atoms have 8 protons) are heavier than hydrogen atoms (with atomic number 1, i.e. a single proton). Oxygen is also the most abundant element in the earth's crust, and is found mainly in water in the human body, as well as being a fundamental component of all proteins, nucleic acids (such as DNA and RNA), carbohydrates and fats.

Although the human body contains more than 60 different types of chemical elements, most are only present in minimal amounts. The human body does not contain helium (a volatile gas with atomic number 2, i.e., the second

Table 5.1 Composition of the human body (average 70 kg/150 lb adult human body)

Element	Atomic number	Atomic percent	Body weight (%)	Mass (kg)
Oxygen	8	24	65	43
Carbon	6	12	18	16
Hydrogen	1	62	10	7.0
Nitrogen	7	1.1	3.0	1.8
Calcium	20	0.22	1.4	1.0
Phosphorus	15	0.22	1.1	0.78
Potassium	19	0.033	0.020	0.14
Sulfur	16	0.038	0.020	0.14
Sodium	11	0.024	0.015	0.095
Chlorine	17	0.037	0.015	0.010
Another 50 elements	3 ~ 92	0.328	1.430	0.035

Source: Authors based on Emsley (2011)

element in the chemical table after hydrogen), but we have "traces" from lithium (atomic number 3) to uranium (atomic number 92).

It is estimated that the universe is composed of about 73% hydrogen atoms and 25% helium atoms. All the other "heavier" atoms (with atomic number from 3 onwards) barely represent the remaining 2% of the known universe. Heavy atoms are believed to have been created as a result of star explosions at the beginning of the universe, so we are really "cosmic dust" or "stardust", as described by the American physicist Carl Sagan in his famous book and television show *Cosmos*.[22]

In short, maintaining basic organisms, as is the case with humans, will be easy and cheap when we know how to repair matter at the atomic and molecular scale, as we are starting to do at the biological level. If we consider nanotechnology as a form of "artificial" biology, it is very likely that we will succeed in repairing atoms in the coming decades.

The idea of atomic and molecular manufacturing was popularized by the American engineer Eric Drexler, who in 1986 published his book *Engines of Creation: The Coming Era of Nanotechnology*. In that work, Drexler, with assistance from MIT artificial intelligence expert Marvin Minsky, formalises the bases of molecular nanotechnology within a project that formed part of his doctoral thesis.[23]

In 2013, Drexler wrote *Radical Abundance*, explaining how impressive advances in nanotechnology will enable us to compose, decompose, and recompose matter at a very low cost, probably as little as a dollar per kilogram. In other words, with advanced nanotechnology, in the next few decades a 70 kg person could be fixed for 70 dollars, and probably for less. If we add all the elements that make up a human, we see that all our chemical elements cost less than 100 dollars in total.

Unless we demand the human body to be filled with Evian or Perrier water from France, the market price of a human being's components is really low. Humans are composed of some the most abundant elements of the earth's crust: we are not made of plutonium (atomic number 94) with diamonds and gold incrustations in antimatter, we are basically made of water, with a little carbon and nitrogen (in addition to "traces" of other elements present in the environment). We are what we consume from the air, the water, and the food that we inhale, drink, and eat.

Thanks to continuous discoveries, biology and medicine keep advancing by leaps and bounds. The infamous medical bloodletting, which was used for centuries and even until the middle of the twentieth century in some parts of the world, is today regarded as little short of barbaric. In a few years we will think the same of the current radiotherapies and chemotherapies. If we exaggerate a little, trying to kill a tumor with radiotherapy or chemotherapy is like killing a mosquito with a cannon. Let's hope that soon radiotherapy and chemotherapy will go down in the history of barbaric methods as the bloodletting of the past did.

To advance in the cure of aging we must think of the basics. The famous South African, Canadian, and American engineer and inventor Elon Musk explains that his success is due to his fixation on basic principles and not by analogy. When we think by analogy, we copy other ideas and only linear improvements are produced. When we think of basic principles, we can visualize exponential changes, up to the limits imposed by science. Musk gives the example of physics as a basic science for reasoning[24]:

I think it's important to reason from first principles rather than by analogy. I do think there's a good framework for thinking. It is physics. Generally, I think there are—what I mean by that is, boil things down to their fundamental truths and reason up from there, as opposed to reasoning by analogy.

Through most of our life, we get through life by reasoning by analogy, which essentially means copying what other people do with slight variations.

Musk continues with the example of batteries for electric cars, and explains why their costs will continue to fall rapidly if we think about the basic principles:

Somebody could say, "Battery packs are really expensive and that's just the way they will always be... Historically, it has cost $600 per kilowatt hour. It's not going to be much better than that in the future."

With first principles, you say, "What are the material constituents of the batteries? What is the stock market value of the material constituents?"

It's got cobalt, nickel, aluminum, carbon, some polymers for separation and a seal can. Break that down on a material basis and say, "If we bought that on the London Metal Exchange what would each of those things cost?"

It's like $80 per kilowatt hour. So clearly you just need to think of clever ways to take those materials and combine them into the shape of a battery cell and you can have batteries that are much, much cheaper than anyone realizes."

This kind of reasoning has allowed Musk to revolutionize the payment industry with PayPal, the solar industry with SolarCity, the electric car industry with Tesla Motors, the space industry with Space X, the transportation industry with Hyperloop, and tunnels with The Boring Company. As if that weren't enough, Musk is working to do the same with brain-computer interfaces through Neuralink, and promoting friendly artificial intelligence through the OpenAI initiative (an open platform for artificial intelligence).[25]

If we focus on the basic fundamentals, we will see that the human body is not so complex, and that we will be able to repair it with new technologies, such as nanotechnology. The human body is also cheap, and fixing something cheap is going to be cheap when we know how to do it well. There will be no bloodletting or chemotherapy or radiotherapy in the future. Today we also know that there are cells and organisms that do not age, and that is the proof of concept that not aging is biologically possible, because it already happens in nature itself. Now we have to understand and replicate non-aging through the basics.

As American futurist Ray Kurzweil explains, all technologies are expensive and bad at first, but become cheaper and better when they become popular. We all know the example of mobile phones. When they went on sale, the first models cost thousands of dollars, they were huge, they worked badly, the batteries ran down quickly, and the devices only served to make or receive calls. Today, thanks to the generalization of technology, mobile phones are cheap and very good, in fact, today they are called smart phones because of the countless tasks they perform through more and better applications, many of them free. Today everyone, everywhere in the world, has a mobile phone if they want to.

At the biotechnological level, sequencing the human genome, which began in 1990 and ended in 2003, is even more impressive. In other words, sequencing the first human genome took 13 years, at a cost of about 3 billion dollars. Thanks to the continuous exponential advances in technology, it is very likely that in a few years the genome can be sequenced for as little as ten dollars in a minute. Devices for sequencing the genome with a connection to the smartphones of the future are likely to be developed soon.

Another example is the human immunodeficiency virus (HIV), which took years to identify and was once considered a "death sentence" by directly attacking the immune system of the infected person. Thanks to the acceleration of technological changes, Kurzweil emphasizes that[26]:

The pace of change is exponential, not linear. So things fifty years from now will be very different. That's pretty phenomenal. It took us fifteen years to sequence HIV, we sequenced SARS in 31 days.

We're doubling the power of computers every year for the same cost. In 25 years, they'll be a billion times more powerful than they are today. At the same time, we're shrinking the size of all technology, electronic and mechanical, by a factor of a hundred per decade, that's a hundred thousand in 25 years.

It took years to identify HIV, it took years to sequence the virus, and it took years to develop the first treatments. The first anti-HIV therapies cost millions of dollars a year, but quickly became widespread and dropped to thousands, then hundreds of dollars a year. In countries like India there are generic treatments for only tens of dollars against HIV. In a few years, we may have only a few dollars in therapies and then get a definitive cure. Today HIV is a chronic controllable disease, not a mortal disease, so is diabetes.

We must aim for the same thing with aging: to make it a controllable chronic disease, and later, to cure it definitively. Thanks to exponential advances it is even possible that we can cure aging before it becomes a chronic disease.

It is essential to begin human trials with rejuvenation technologies that have proven useful in other animals. This is one of the objectives of the Project 21 of the SENS Research Foundation.[27]

In order to cure aging, we must focus investments on the causes and thus avoid spending on the symptoms. In an interview, the English biogerontologist Aubrey de Grey commented that 90% of deaths and at least 80% of medical costs in the United States are due to aging. However, a tiny amount of resources is devoted to it, and there is not enough that foundations like SENS can do for themselves if they do not have more public or private support. As an example, let's compare the following budgets[28]:

National Institutes of Health budget ($M): ~30,000
National Institute of Aging budget: ~1,000
Division of Aging Biology budget: ~150
Spent on translational research (max): ~10
SENS Research Foundation budget: ~5

Again, let's remember that global medical spending is around $7 trillion a year, every year, and rising. Unfortunately, almost all spending goes to the last years of life, and without much success, as in the end patients die as well, often in tragic conditions. We must rethink the entire health system and invest at the beginning, rather than spend at the end. As the proverb goes, "prevention is better than cure".

To move in the right direction, we must also change our own mentality and accept death as a terrible enemy, the greatest enemy of all humanity, but an enemy we can defeat. If we abandon the fear of death and act with our brains and hearts, then we will come to the death of death.

Notes

1. https://www.theguardian.com/world/2008/may/27/japan
2. http://www.theguardian.com/world/2013/jan/22/elderly-hurry-up-die-japanese
3. http://www.nytimes.com/1984/03/29/us/gov-lamm-asserts-elderly-if-very-ill-have-duty-to-die.html
4. http://www.theatlantic.com/magazine/archive/2014/10/why-i-hope-to-die-at-75/379329/
5. http://www.amazon.com/Reinventing-American-Health-Care-Outrageously/dp/1610393457
6. https://web.archive.org/web/20110110154034/http:/www.sagecrossroads.net/files/transcript01.pdf
7. http://www.ncbi.nlm.nih.gov/pmc/articles/PMC1361028/
8. http://sjayolshansky.com/sjo/Background_files/TheScientist.pdf
9. http://scholar.harvard.edu/cutler/publications/substantial-health-and-economic-returns-delayed-aging-may-warrant-new-focus
10. http://www.reuters.com/article/us-imf-aging-idUSBRE83A1C020120412
11. http://www.brookings.edu/research/books/2013/closing-the-deficit
12. http://articles.latimes.com/2014/jan/08/business/la-fi-mo-sure-you-have-to-work-in-retirement-but-look-on-the-bright-side-20140108
13. http://www.nber.org/papers/w8818
14. https://web.archive.org/web/20061018172529/http://www.econ.yale.edu/seminars/labor/lap04-05/topel032505.pdf
15. https://web.archive.org/web/20150908111708/http://www.northbaybusinessjournal.com/northbay/marincounty/4138872-181/quest-to-redefine-aging#page=0
16. httpm//www.nature.com/articles/s43587-021-00080-0#auth-Martin-Ellison
17. https://report.nih.gov/categorical_spending.aspx

18. http://www.forbes.com/sites/alexknapp/2012/07/05/how-much-does-it-cost-to-find-a-higgs-boson/
19. http://waterwaysproducts.com.au/2017/03/water-affect-human-body/
20. https://web.archive.org/web/20171019041147/https://www.nestle-waters.com/healthy-hydration/water-body
21. https://www.amazon.com/Natures-Building-Blocks-Z-Elements/dp/0199605637
22. https://www.amazon.com/Carl-Sagan-Cosmos-Utimate-Blu-ray/dp/B06X1F546N
23. https://www.amazon.com/Engines-Creation-Nanotechnology-Library-Science/dp/0385199732/
24. https://www.youtube.com/watch?v=NV3sBlRgzTI
25. https://www.supplychaintoday.com/openai-neuralink-shaping-our-ai-future/
26. http://edition.cnn.com/2006/TECH/science/06/12/introduction/
27. http://sensproject21.org/
28. https://medium.com/@arielf/wake-up-people-its-time-to-aim-high-b0c2bcac53f1

6

The Terror of Death

Man fears death because he loves life.
Fyodor Dostoyevsky, 1880

All great truths begin as blasphemies.
George Bernard Shaw, 1919

I don't believe in the afterlife, although I am bringing a change of underwear.
I am not afraid of death, I just don't want to be there when it happens.
Woody Allen, 1971

Technology is accelerating. As a consequence, "rejuveneering" – the engineering of rejuvenation – is poised to jump ahead in leaps and bounds. The waves of improvements to society that arose in the wake of the industrial revolution – the growth of the economy, better education, greater mobility, improved healthcare, richer opportunities – are set to continue, at an even more rapid pace than before. Ever larger numbers of people are both willing and capable to become involved in research and development activities, in multiple converging technology sectors, as part of a huge extended global network:

- More engineers, scientists, designers, analysts, entrepreneurs, and other change agents are being trained now, at universities and elsewhere, than ever before
- The availability of high-quality online education material, often free-of-charge, means that these budding technologists are starting from a higher base point than most of their predecessors of just a few years ago
- People at later stages in their professional lives are able to jump into fertile new fields – perhaps initially "only browsing" – using some of their discretionary free time; this applies in particular to people who have retired from

J. Cordeiro, D. Wood, *The Death of Death*, Copernicus Books,
https://doi.org/10.1007/978-3-031-28927-9_6

their previous work, or who have been made redundant but who still have many skills that they can deploy
- Connections between these different researchers, via a myriad of online communications channels, wikis, databases, AI linkages, and so on, mean that bright researchers can more quickly find out about promising lines of analysis happening elsewhere in the world
- The increasing prevalence of open source software, freely distributed, further helps to encourage wider participation.

It's this positive network effect – more people, better educated, better networked, building on top of each other's solutions – that leads the authors to the conclusion that, other things remaining equal, the overall pace of technology improvement is likely to rise and rise. The rapid breakthroughs of the last few decades in IT, smartphones, 3D printing, genetic engineering, brain scanning, and so on, are likely to be matched (if not surpassed) by similarly rapid breakthroughs in numerous other fields in the next few decades. Of critical importance, this pattern applies to innovation in medical treatments, and in particular, to innovation in rejuveneering.

Of course, there exist many tough obstacles which could impede progress in potential medical breakthroughs. This includes regulatory hurdles and other examples of system complexity and system inertia. Nevertheless, there are also an unprecedented number of well-educated, capable people who are already busy exploring possible solutions and workarounds for these hurdles. In a spirit of "divide and conquer", they're working on improved tools, libraries, test modules, methodologies, alternative regulatory pathways, AI analysis of big medical data, and much, much more. They can build creatively on each other's insights. And when they achieve good results, larger companies can join forces with them, to give their ideas any extra push needed.

Early signs of accomplishments by rejuveneering are all around us. The field can no longer be dismissed (as it used to be, by some critics) as quack medicine and snake oil. A host of interesting lines of inquiry await further research and development. Some of these lines of enquiry may turn out to be fruitless, but there's no reason to think that the field in its entirety will remain barren.

What's more, there are powerful economic reasons for continuing this work. People who benefit from rejuveneering will, other things being equal, contribute more to the economy, and to the overall social capital. It makes strong financial sense for society to accelerate its investment in rejuveneering.

If there's a strong financial case for something happening, you'd think that society ought to be able to agree and say, "Let's do it". But that's far from being the case with rejuveneering. Instead, there are layers of opposition to it. It's time to dig more deeply into the roots of that opposition.

Varieties of Objection

People often have a set of objections to the rejuveneering project. Some of the most frequently cited ones are:

- How does rejuveneering propose to address incurable diseases?
- Don't principles of physics, such as entropy, render rejuveneering impossible?
- Isn't the rejuveneering programme so inherently complicated it will require centuries of work?
- Aren't there natural limits to how long humans can live?
- Won't rejuveneering cause a dreadful population explosion?
- Won't long-lived people provide a brake on necessary societal change?
- In the absence of aging and death, what motivation will people have to get anything done?
- Won't the wealthy benefit disproportionately from rejuveneering?
- Isn't it egocentric to pursue rejuveneering?

In each case, rejuveneers have a strong answer. We give many good answers throughout this book. Nevertheless, these good answers, by themselves, seem insufficient to change the minds of critics and sceptics. There's something deeper going on.

To understand what's happening, we need to distinguish *underlying motivation* from *supportive rationale*. In the vivid metaphor of the elephant and the rider, developed by social psychologist Jonathan Haidt in his book *The Happiness Hypothesis*, the conscious mind is akin to the human rider bestride a powerful elephant – the subconscious. Haidt develops this analogy in the first chapter of his book[1]:

> Why do people keep doing… stupid things? Why do they fail to control them-selves and continue to do what they know is not good for them? I, for one, can easily muster the willpower to ignore all the desserts on the menu. But if dessert is placed on the table, I can't resist it. I can resolve to focus on a task and not get up until it is done, yet somehow I find myself walking into the kitchen, or pro-

crastinating in other ways. I can resolve to wake up at 6:00 A.M. to write; yet after I have shut off the alarm, my repeated commands to myself to get out of bed have no effect…

It was during some larger life decisions, about dating, that I really began to grasp the extent of my powerlessness. I would know exactly what I should do, yet, even as I was telling my friends that I would do it, a part of me was dimly aware that I was not going to. Feelings of guilt, lust, or fear were often stronger than reasoning…

Modern theories about rational choice and information processing don't adequately explain weakness of the will. The older metaphors about controlling animals work beautifully. The image that I came up with for myself, as I marvelled at my weakness, was that I was a rider on the back of an elephant. I'm holding the reins in my hands, and by pulling one way or the other I can tell the elephant to turn, to stop, or to go. I can direct things, but only when the elephant doesn't have desires of his own. When the elephant really wants to do something, I'm no match for him.

The rider may think he or she is in control, but the elephant often has its own firm ideas, particularly in matters of taste and morality. In such cases, the conscious mind acts more like a lawyer than a driver. As Haidt continues:

Moral judgment is like aesthetic judgment. When you see a painting, you usually know instantly and automatically whether you like it. If someone asks you to explain your judgment, you confabulate. You don't really know why you think something is beautiful, but your interpreter module (the rider) is skilled at making up reasons… You search for a plausible reason for liking the painting, and you latch on to the first reason that makes sense (maybe something vague about colour, or light, or the reflection of the painter in the clown's shiny nose). Moral arguments are much the same: Two people feel strongly about an issue, their feelings come first, and their reasons are invented on the fly, to throw at each other. When you refute a person's argument, does she generally change her mind and agree with you? Of course not, because the argument you defeated was not the cause of her position; it was made up after the judgment was already made.

If you listen closely to moral arguments, you can sometimes hear something surprising: that it is really the elephant holding the reins, guiding the rider. It is the elephant who decides what is good or bad, beautiful, or ugly. Gut feelings, intuitions, and snap judgments happen constantly and automatically… but only the rider can string sentences together and create arguments to give to other people. In moral arguments, the rider goes beyond being just an advisor to the elephant, he becomes a lawyer, fighting in the court of public opinion to persuade others of the elephant's point of view.

In his follow-up book, *The Righteous Mind*, Haidt builds upon that metaphor to propose a cornerstone principle of moral psychology: intuitions come
first, strategic reasoning second[2]:

> Moral intuitions arise automatically and almost instantaneously, long before
> moral reasoning has a chance to get started, and those first intuitions tend to
> drive our later reasoning. If you think that moral reasoning is something we do
> to figure out the truth, you'll be constantly frustrated by how foolish, biased,
> and illogical people become when they disagree with you. But if you think
> about moral reasoning as a skill we humans evolved to further our social agen-
> das—to justify our own actions and to defend the teams we belong to—then
> things will make a lot more sense. Keep your eye on the intuitions, and don't
> take people's moral arguments at face value. They're mostly post-hoc construc-
> tions made up on the fly, crafted to advance one or more strategic objectives.
>
> The central metaphor… is that *the mind is divided, like a rider on an elephant,
> and the rider's job is to serve the elephant.* The rider is our conscious reasoning—
> the stream of words and images that hogs the stage of our awareness. The ele-
> phant is the other 99% of mental processes—the ones that occur outside of
> awareness but that actually govern most of our behavior.

The biggest challenge facing the rejuveneering project isn't the set of supportive rationales that critics bring forward, as reasons for their opposition to
the project. Instead, what urgently needs re-tuning is the underlying motivation that's guiding these critics, often without their conscious awareness. It's
not the rider that we need to argue with. Instead, we have to find ways of
directly engaging the elephant.

Managing Terror

It's a fundamental fact of animals that they can experience terror. When facing
a tangible threat of death, an animal's metabolism jumps into a different gear.
Glands produce the hormones adrenaline and cortisol, which speed up heart-
beat, dilate the pupils in the eye to take in more information about the
impending danger, and increase blood flow to muscles and lungs in prepara-
tion for violent action. The animal is primed for either fight or flight. So that
maximum energy is available for urgent self-preservation activities, other
bodily processing is slowed down, including the digestion of food. Peripheral
vision is reduced, so that the animal can concentrate more fully on the immediate threat at hand. Loss of hearing occurs too.

Terror is a state that serves a vital purpose when, indeed, the animal is under imminent mortal threat. In that state, the body is optimised to survive the immediate challenge. However, that state is far from optimised for longer-term existence. On the contrary, when in a state of panic, attention is restricted, thought patterns narrow, digestion suffers, and the body can be overtaken by convulsion and shaking. Uncontrollably releasing the contents of your bladder and sphincter may have the benefit of, perhaps, disgusting and repulsing would-be attackers, but it's not conducive to healthy social living at other times.

The human ability to vividly anticipate death ahead of time – that is, when not in any imminent danger – poses a problem for the management of the body's terror subsystem. If the thought of death becomes all-consuming, normal processing becomes impossible. Worse, another aspect of animal psychology is that terror is contagious: if one animal in a group has spotted a predator nearby, the whole group can react quickly and decisively. Likewise, if one human becomes panic-stricken, their mood can swiftly spread, even in the absence of an objective cause for any panic.

The management of terror is, therefore, a key problem for human society. This has been the case right back into early prehistory, when humans started to acquire the capacities for self-awareness, planning, and introspective reflection. Observing the increasing frailty of group members who had, at younger times, been marvellously fit and healthy, early humans would be struck by the thought that a similar decline awaited them – and everyone else who they loved and cherished. In other words, mortal terror switched from being an occasional state, necessary for individual survival, to being something that could well up in someone's mind at any moment, unbidden by any external threat, leaving the person paralysed and panic-stricken.

What's more, the conscious anticipation of death from threats such as predators or competing bands of humans would tend, other things being equal, to cause strong risk-aversion. Behaviour that reduces short-term risk – such as remaining hidden in the depths of a cave – may well be far from the best for the long-term progress of the group.

For these reasons, we can reasonably speculate that the groups of humans who successfully survived tended to be ones that developed social and psychological tools to manage the terror of the prospect of death – terror that would otherwise incapacitate the group. These tools in various ways denied the awfulness of the threat of death. These tools included mythology, tribalism, religion, ecstatic trances, and the appearance of contact with spirits. In later times, these tools also included cultural mores and thinking patterns which held out the promise of different kinds of transcendence of physical death, via the survival of our legacy, or the survival of a larger group of which we form an integral part. These thinking patterns are tied up with elements of our social philosophy – the way that we conceive of who we are, how we fit into our society, and how our society fits into the larger cosmos.

Our social philosophy, therefore, provides an important element of mental stability against the ever-lurking existential dread of mortality. But this means that anything which challenges our social philosophy – anything that suggests our philosophy has major flaws – is itself a danger to our mental well-being. Sensing this, our inner elephant can go wild, leading us to all sorts of irrational behaviour – behaviour which our inner lawyer/rider then hurries to rationalise.

What we have just described is a theory that was popularised by the philosopher Ernest Becker in his 1973 Pulitzer Prize-winning book *The Denial of Death*.[3]

Beyond the Denial of Death

Becker wrote the following words at the start of *The Denial of Death*:

> The prospect of death, Dr. Johnson said, wonderfully concentrates the mind. The main thesis of this book is that it does much more than that: the idea of death, the fear of it, haunts the human animal like nothing else; it is a mainspring of human activity—activity designed largely to avoid the fatality of death, to overcome it by denying in some way that it is the final destiny for man.

Sam Keen, contributing editor at Psychology Today, contributed a foreword to *The Denial of Death*, in which he described Becker's philosophy as being "a braid woven from four strands":

- The world is terrifying
- The basic motivation for human behaviour is our biological need to control our basic anxiety, to deny the terror of death
- Since the terror of death is so overwhelming we conspire to keep it unconscious
- Our heroic projects that are aimed at destroying evil have the paradoxical effect of bringing more evil into the world.

Becker's thesis is sweeping. It's one of a handful of ideas that attempt to show that human history has been shaped by forces that we often prefer not to acknowledge:

- Galileo defended that the Earth is not the center of the universe but just another small planet.
- Darwin showed that humans do not descend from gods but from other inferior apes.
- Marx highlighted the role of class conflict and social alienation
- Freud highlighted repressed sexuality
- Becker highlighted our desire to deny the reality of death.

In common with all large theories of this form, Becker's thesis has critics who ask: where is the evidence? Sadly, Becker himself was unable to directly respond to such critics, as he had died even before the publication of *The Denial of Death*, struck down by colon cancer. Sam Keen includes in his foreword a poignant account of him meeting Becker, for the first time, whilst Becker was at death's door:

> The first words Ernest Becker said to me when I walked into his hospital room were: "You are catching me in extremis. This is a test of everything I've written about death. And I've got a chance to show how one dies, the attitude one takes. Whether one does it in a dignified, manly way; what kinds of thoughts one surrounds it with; how one accepts his death…"
>
> Although we had never met, Ernest and I fell immediately into deep conversation. The nearness of his death and the severe limits of his energy stripped away the impulse to chatter. We talked about death in the face of death; about evil in the presence of cancer. At the end of the day Ernest had no more energy, so there was no more time. We lingered awkwardly for a few minutes, because saying "goodbye" for the last time is hard and we both knew he would not live to see our conversation in print. A paper cup of medicinal sherry on the night stand, mercifully, provided us a ritual for ending. We drank the wine together and I left.

Nevertheless, other researchers have stepped in, to provide a swathe of empirical evidence that fleshes out Becker's theory – evidence from a field sometimes called "experimental existential psychology". This new work was comprehensively summarised in the 2015 book *The Worm at the Core*[4] by social psychologists Jeff Greenberg, Tom Pyszczynski, and Sheldon Solomon.

The phrase used for the title of the book, "The worm at the core", comes from an extract from the 1902 publication *The varieties of religious experience*:

a study in human nature by philosopher William James. The authors quote that extract approvingly, and comment[5]:

> There is now compelling evidence that, as William James suggested a century ago, death is indeed the worm at the core of the human condition. The awareness that we humans will die has a profound and pervasive effect on our thoughts, feelings, and behaviours in almost every domain of human life—whether we are conscious of it or not.
>
> Over the course of human history, the terror of death has guided the development of art, religion, language, economics, and science. It raised the pyramids in Egypt and razed the Twin Towers in Manhattan. It contributes to conflicts around the globe. At a more personal level, recognition of our mortality leads us to love fancy cars, tan ourselves to an unhealthy crisp, max out our credit cards, drive like lunatics, itch for a fight with a perceived enemy, and crave fame, however ephemeral, even if we have to drink yak urine on *Survivor* to get it.

Terror Management Theory

Greenberg, Pyszczynski, and Solomon have coined the acronym TMT as shorthand for "Terror Management Theory" – which is their development of the ideas of Ernest Becker. The Ernest Becker Foundation website carries a description of TMT[6]:

> TMT posits that while humans share with all life-forms a biological predisposition toward self-preservation in the service of reproduction, we are unique in our capacity for symbolic thought, which fosters self-awareness and the ability to reflect on the past and ponder the future. This spawns the realization that death is inevitable and can occur at any time for reasons that cannot be anticipated or controlled.
>
> The awareness of death engenders potentially debilitating terror that is "managed" by the development and maintenance of cultural worldviews: humanly constructed beliefs about reality shared by individuals that minimize existential dread by conferring meaning and value. All cultures provide a sense that life is meaningful by offering an account of the origin of the universe, prescriptions for appropriate behaviour, and assurance of immortality for those who behave in accordance with cultural dictates. Literal immortality is afforded by souls, heavens, afterlives, and reincarnations associated with all major religions. Symbolic immortality is obtained by being part of a great nation, amassing great fortunes, noteworthy accomplishments, and having children.
>
> Psychological equanimity also requires that individuals perceive themselves as persons of value in a world of meaning. This is accomplished through social

roles with associated standards. Self-esteem is the sense of personal significance that results from meeting or exceeding such standards.

The website also summarises three lines of the empirical evidence that support TMT:

1. The anxiety-buffering function of self-esteem is established by studies where momentarily elevated self-esteem results in lower self-reported anxiety and physiological arousal.
2. Making death salient by asking people to think about themselves dying (or viewing graphic depictions of death, being interviewed in front of a funeral parlour, or subliminal exposure to the word "dead" or "death") intensifies strivings to defend their cultural worldviews by increasing positive reactions to similar others, and negative reactions toward those who are different.
3. Research verifies the existential function of cultural worldviews and self-esteem by demonstrating that non-conscious death thoughts come more readily to mind when cherished cultural beliefs or self-esteem is threatened.

TMT has generated empirical research (currently more than 500 studies) examining a host of other forms of human social behaviour, including aggression, stereotyping, needs for structure and meaning, depression and psychopathology, political preferences, creativity, sexuality, romantic and interpersonal attachment, self-awareness, unconscious cognition, martyrdom, religion, group identification, disgust, human-nature relations, physical health, risk taking, and legal judgments.

In summary, there are deep roots to the opposition that many people have to the idea of extending healthy lifespans. People may offer intellectual rationalisations for their opposition (e.g. "How would humankind cope with tens of millions of extremely old and incredibly crabby people?") but these rationalisations are not the drivers for the position they hold.

Instead, their opposition to extending healthy lifespans comes from what we can call faith. Thomas Pyszczynski of the University of Colorado explained this attitude in a presentation for the SENS6 conference entitled "Understanding the paradox of opposition to long-term extension of the human lifespan: fear of death, cultural worldviews, and the illusion of objectivity".[7]

The Paradox of Opposition to Extended Healthspan

Here's the "paradox" to which Pyszczynski referred in the title of his talk: Nobody wants to die, but many people object to long-term extension of the human lifespan by reversing the aging process. Pyszczynski's explanation is that it's the operation of an entrenched "anxiety buffering system" – a mix of culture and philosophy – which leads people to oppose the idea that we could have longer healthy lives. This anxiety buffering system was originally an adaptive response to the disturbing underlying fact that something we deeply desire – indefinitely long healthy lives – is unachievable.

For all of history up until the present age, the aspiration to have an indefinitely long healthy life was at stark variance to everything else that we saw around ourselves. Death seemed inevitable. To reduce the risk of collapsing into terror at this realisation, we needed to develop rationalisations and techniques that prevented us from thinking counterproductively about our own finitude and mortality. That's where key aspects of our culture arose, creating and sustaining our elaborate anxiety buffering system. Meeting an important social need, these aspects of our culture became deeply rooted.

Our culture often operates below the level of conscious awareness. We find ourselves being driven by various underlying beliefs, without being aware of the set of causes and effects. However, we find comfort in these beliefs, especially when "other people like us" also espouse these beliefs, providing a measure of social validation. This faith (belief in the absence of sufficient reason) helps to keep us mentally sane, and keeps society functional, even as it prepares us, as individuals, to grow infirm and die.

To be clear, the "faith" described here, as intrinsic to the continuation of the accepting-aging paradigm, may or may not involve (for any specific individual) a belief in a supernatural "life beyond death", such as many religions describe. But the faith does, in all cases, involve the view that good members of society should accept death when their time comes, that society could not function properly if individuals ignored that principle, and that the fundamental meaning of an individual's life is tied up in the longer-term flourishing of the society or tradition of which they are part.

In case any new ideas challenge this faith, adherents often find themselves compelled to lash out against these ideas, even without taking the time to analyse them. Their motivation is to preserve their core culture and faith, since that's what provides the foundation of meaning in their lives. They fight the new ideas, even if these new ideas would be a better solution to their underlying desire to live an indefinitely long, healthy life. Paradoxically, it's their fear of death that makes them upset about the contrary ideas. These ideas generate feelings of alienation, even though they don't see the actual mental connections between the ideas. In summary, their faith causes them to lose their rationality.

Another useful metaphor here, Pyszczynski suggests, is to see our anxiety buffering system as a psychological immune system that seeks to destroy incoming ideas which would cause us mental distress. Like our physical immune system, our psychological immune system sometimes malfunctions, and attacks something that would actually bring us greater health.

Aubrey de Grey has also written on this topic. In chapter two of his 2007 book *Ending Aging*, he notes the following[8]:

> There is a very simple reason why so many people defend aging so strongly – a reason that is now invalid, but until quite recently was entirely reasonable. Until recently, no one has had any coherent idea how to defeat aging, so it has been effectively inevitable. And when one is faced with a fate that is as ghastly as aging and about which one can do absolutely nothing, either for oneself or even for others, it makes perfect psychological sense to put it out of one's mind – to make one's peace with it, you might say – rather than to spend one's miserably short life preoccupied by it. The fact that, in order to sustain this state of mind, one has to abandon all semblance of rationality on the subject – and, inevitably, to engage in embarrassingly unreasonable conversational tactics to shore up that irrationality – is a small price to pay…

In his analysis, de Grey refers to "the pro-aging trance", on account of what he describes as "the depth of irrationality that is exhibited by so many people".[9] Other writers refer to the concept of "deathism"; for example, the website "Fight Aging!" has published what it called "An Anti-Deathist FAQ".[10] The term "Accepting-aging paradigm" is preferable, since it is less pejorative and may lower the temperature of what can already be a heated discussion.

Engaging the Elephant

Let's return to excellent advice offered by Jonathan Haidt, regarding changing the direction of the "elephant" which represents our subconscious tendencies. If we recognise that these tendencies are flawed, as in the case of the accepting-aging paradigm, what can we do to change them? The following is from the third chapter, "Elephants Rule", of his book *The Righteous Mind*[11]:

The elephant is far more powerful than the rider, but it is not an absolute dictator. When does the elephant listen to reason? The main way that we change our minds on moral issues is by interacting with other people. We are terrible at seeking evidence that challenges our own beliefs, but other people do us this favour, just as we are quite good at finding errors in other people's beliefs. When discussions are hostile, the odds of change are slight. The elephant leans away from the opponent, and the rider works frantically to rebut the opponent's charges. But if there is affection, admiration, or a desire to please the other person, then the elephant leans toward that person and the rider tries to find the truth in the other person's arguments. The elephant may not often change its direction in response to objections from its own rider, but it is easily steered by the mere presence of friendly elephants or by good arguments given to it by the riders of those friendly elephants…

Under normal circumstances the rider takes its cue from the elephant, just as a lawyer takes instructions from a client. But if you force the two to sit around and chat for a few minutes, the elephant actually opens up to advice from the rider and arguments from outside sources. Intuitions come first, and under normal circumstances they cause us to engage in socially strategic reasoning, but there are ways to make the relationship more of a two-way street…

The elephant (automatic processes) is where most of the action is in moral psychology. Reasoning matters, of course, particularly between people, and particularly when reasons trigger new intuitions. Elephants rule, but they are neither dumb nor despotic. Intuitions can be shaped by reasoning, especially when reasons are embedded in a friendly conversation or an emotionally compelling novel, movie, or news story.

This provides us with three ways to change the elephant's opinion, on a matter as controversial as whether healthy life extension is ultimately a desirable or an undesirable outcome. People are more likely to accept advice, on potentially difficult topics, if that advice:

1. Comes from someone perceived as being "one of us" – that is, a friend, from a similar demographic, rather than being a strange outsider
2. Is supported by "an emotionally compelling novel, movie, or news story"
3. Exists in a context where the elephant feels that its own needs are well understood and well supported.

The first of these conditions matches a well-known principle of technology marketing – that of companies needing to change their marketing approach while "crossing the chasm" from the set of early adopters of a new technology, to the larger market of "early majority". Geoffrey Moore brought attention to this idea in his 1991 book *Crossing the Chasm*,[12] which in turn drew on rich observations by Everett Rogers' 1962 work *The Diffusion of Innovation*.[13] The key insight is as follows: whereas early adopters of a new idea are prepared to act as visionaries, access to the mainstream market is controlled by pragmatists whose strong instinct is to "stick with the herd". In general, such people will only adopt a solution (or an idea) if they see others from their own herd who have already adopted it and endorse it.

There's an important implication here. Advocates and slogans that were successful in attracting an initial community of supporters to a new cause – such as the anticipating-rejuvenation paradigm – often need to be changed, before potential mainstream supporters will be prepared to listen. Talk of, for example, immortality, or mind uploading, which appealed to early supporters of rejuveneering, can be counterproductive as the movement seeks to gain a wider circle of supporters. People who might endorse the longevity dividend might be repelled by talk of the defeat of death.

The second and third of the above three conditions were addressed in a presentation by another speaker at the aforementioned SENS6. This speaker was Mair Underwood of the University of Queensland. Her presentation was entitled "What reassurances do the community need regarding life extension? Evidence from studies of community attitudes and an analysis of film portrayals".[14]

Underwood's presentation pointed out the many ways in which would-be rejuveneers are depicted in a bad light in popular films including *The Fountain*, *Death Becomes Her*, *Highlander*, *Interview with the Vampire*, *Vanilla Sky*, *Dorian Gray*, and so on. Rejuveneers, these films imply, are emotionally immature, selfish, reckless, obstructive, narrow-minded, and generally dislikeable. The heroes in these films – the characters who are portrayed as calm, rational, praiseworthy, and mentally healthy – are the characters that voluntarily choose *not* to extend their lives.

Films with a contrary, affirmative impression of life extension are much less common; *Cocoon*, directed by Ron Howard, is perhaps the best-known example. One reason that negative stereotypes prevail in popular films is, no doubt, because dystopia tends to sell better than utopia. However, Hollywood stereotypes draw their strength from pre-existing cultural norms. As such, these films reflect and magnify viewpoints about life extension that are already widely distributed among the general population:

- Life extension would be boring and repetitive
- Long term relationships would suffer
- Life extension would mean the extension of chronic illness
- Life extension would be unfairly distributed.

To counteract these negative viewpoints, and to help free society from its accepting-aging paradigm, Underwood gave the following advice to the rejuveneering community:

1. Avoid berating the general public for having "breath-taking stupidity" on the subject of life extension
2. Provide assurances that life extension science, and the distribution of life extension technologies, are ethical and regulated, and seen to be so
3. Assuage community concerns about life extension as "unnatural" or "playing god"
4. Provide assurances that life extension would involve an extension of healthy lifespan
5. Provide assurances that life extension does not mean a loss of sexuality or fertility
6. Provide assurances that life extension will not exacerbate social divides, and that those with extended lives will not be a burden on society
7. Create a new cultural framework for understanding life extension.

We've been aiming to follow that advice throughout this book. It is necessary to convey positive visions about the kind of society that can emerge from the awakening of the engineering of rejuvenation: a new cultural framework to understand not only the extension of life, but also the expansion of life. To advance towards the death of death, we must first leave behind the terror of death itself.

Notes

1. http://www.happinesshypothesis.com/happiness-hypothesis-ch1.pdf
2. http://righteousmind.com/about-the-book/introductory-chapter/
3. http://www.amazon.com/Denial-Death-Ernest-Becker/dp/0684832402/
4. http://www.amazon.com/Worm-Core-Role-Death-Life/dp/1400067472/
5. http://www.amazon.com/Varieties-Religious-Experience-William-James/dp/1482738295/
6. http://ernestbecker.org/?page_id=60
7. https://www.youtube.com/watch?v=biNF_a5QbwE
8. http://www.amazon.com/Ending-Aging-Rejuvenation-Breakthroughs-Lifetime/dp/0312367074/
9. https://www.youtube.com/watch?v=RITCdrOEO9Y
10. https://www.fightaging.org/archives/2014/07/an-anti-deathist-faq.php

11. http://www.amazon.com/Righteous-Mind-Divided-Politics-Religion/dp/0307455777/
12. http://www.amazon.com/gp/product/0062292986/
13. http://www.amazon.com/Diffusion-Innovations-5th-Everett-Rogers/dp/0743222091/
14. https://www.youtube.com/watch?v=vg4lTZvfIz8

7

"Good", "Bad" and "Expert" Paradigms

> I intend to live forever, or die trying.
> **Groucho Marx, 1960**

> Why was I born if it wasn't forever?
> **Eugène Ionesco, 1962**

> Scientific views end in awe and mystery, lost at the edge in uncertainty, but they appear to be so deep and so impressive that the theory that it is all arranged as a stage for God to watch man's struggle for good and evil seems inadequate.
> **Richard Feynman, 1963**

For debates where both sides have deep mental and social roots, it can be hard work to change opinions. That's certainly the case with the debate whether to accept the inevitability of aging, or instead to embrace the possibility to create a "humanity+" society free from aging. But we can find some encouragement – and draw some lessons – from examples of similar seemingly intractable debates which did, in the end, move forwards.

Optical Illusions and Mental Paradigms

We're all familiar with visual illusions which can be perceived in two different ways. For example, a picture can be either a duck or a rabbit,[1] depending on how we look at it. Another picture can represent either a vase, or two faces looking at each other.[2] Yet another picture – this time, one that animates – can, disturbingly, be seen either as a ballerina rotating clockwise, or as the same ballerina rotating anti-clockwise.[3] In all these examples, what's impossible is to accept both viewpoints simultaneously. Our brains can jump from one perspective to the other, but cannot hold both at once.

© The Author(s), under exclusive license to Springer Nature Switzerland AG 2023
J. Cordeiro, D. Wood, *The Death of Death*, Copernicus Books,
https://doi.org/10.1007/978-3-031-28927-9_7

Something similar happens on occasion in the progress of science, although in that case, the effort to move from one perspective to the other can be even harder. Here, the two conflicting perspectives are two different scientific theories in a given field of knowledge. For example, consider the clash in the sixteenth century between the prevailing Aristotelian principle that bodies left to themselves would tend to come to rest, and the new idea, championed by Galileo, that the natural state of affairs was for bodies to continue travelling in straight lines at constant speed. Or consider the clash in the twentieth century between the once-dominant theory that continents have been fixed in place throughout the history of the earth, and the rival new theory that South America and Africa once jostled next to each other, in a long-ago reconfiguration of the continents, before that supercontinent broke up and individual continents drifted apart.

We'll shortly look at some examples of clashing scientific paradigms within the medical sphere. We will also examine the clash between the "accepting aging" paradigm and the rival "anticipating rejuvenation" paradigm. But first, let's look more closely at the intriguing and illuminating case of the theory of continental drift. The hostility shown by mainstream geologists towards the "too large, too unifying, too ambitious" theory of continental drift seemed, at the time, to be amply justified. That fact should give modern-day critics of rejuveneering reason to pause, before they dismiss that theory as being (likewise) "out of the question".

Scientific Hostility

Which child growing up in the twentieth century, looking at a world map, did not wonder to themselves about the similarity of the outlines of South America and Africa? Could these two giant continents once have been part of an even larger whole, somehow split asunder? The same naïve imagination could also chuckle at the thought that, in a similar way, the eastern coastline of North America broadly matched the western coastlines of Northern Africa and Europe. Was this a strange coincidence, or an indication of something more profound?

Mainstream geologists resisted such an idea. To them, the earth was fixed and solid. Ideas to the contrary could be held by naïve school children, but not (they said) by serious scientists.

Even when contrarian writers like Alfred Wegener (from 1912 onwards) and Alex du Toit (from 1937) assembled more data supportive of the idea that continents must have somehow drifted apart from a prehistoric unified landmass, orthodoxy shrugged off the evidence. Wegener and du Toit pointed to surprising similarities of fossils of flora and fauna along the different edges of continents which were now far apart, but which (they suggested) must have been adjacent in bygone times. Moreover, even the rock strata at the edges of these continents matched in surprising ways; for example, rocks in parts of Ireland and Scotland are very similar to those in New Brunswick and Newfoundland, Canada.

But Wegener was an outsider. His doctorate was in astronomy, and his profession was meteorology (weather forecasting). He had no specialist background in geology. Who was he to upturn conventional thinking? Indeed, his lecturing position, at Marburg University, was unpaid – this was seen as another sign that he lacked authority. Wegener's detractors found plenty to criticise:

- Careful cardboard cut-outs of the edges of continents showed that the alleged match was far from snug; the coincidence was by no means as compelling as per first glance
- Wegener's background as an Artic explorer and a high-flying balloonist resulted in jibes that he suffered from "wandering pole plague" as well as "moving crust disease"
- There was no clear mechanism for how continents could actually drift, as part of an earth that was assumed to be solid throughout.

Rollin Chamberlin, an orthodox geologist from the University of Chicago, thundered at a meeting of the American Association of Petroleum Geologists in New York in 1926 that:[4]

If we are to believe Wegener's hypothesis we must forget everything which has been learned in the last 70 years and start all over again.

At the same meeting, Chester Longwell, a geologist from Yale University, exclaimed that:

We insist on testing this hypothesis with exceptional severity, for its acceptance would mean the discarding of theories held so long that they have become almost an integral part of our science.

Richard Conniff, writing in the Smithsonian magazine in an article entitled "When Continental Drift Was Considered Pseudoscience" noted that, for decades afterwards:[5]

Older geologists warned newcomers that any hint of an interest in continental drift would doom their careers.

Eminent English statistician and geophysicist Harold Jeffreys (professor at the University of Cambridge) was another strong opponent of the theory of continental drift. His view was that continental drift was "out of the question", since no force could be sufficient to move continental slabs over the surface of the globe. This was no idle surmise. As explained on the biographical page for Jeffreys on the Penn State University website, Jeffreys had extensive calculations to back up his opinions:[6]

> His main issue with the theory was Wegener's idea about how the continents moved. Wegener stated that the continents simply ploughed through the oceanic crust when they moved. Jeffreys calculated that the Earth is simply too rigid for that to have happened. According to Jeffreys' calculations that if the Earth was weak enough for plates to move through the oceanic crust then mountains would crumble under their own weight.

Wegener also stated that the continents moved in a westward direction due to tidal forces affecting the interior of the planet. Again, Jeffreys' calculations show that tidal forces that strong would stop the rotation of the Earth within a year. Basically, according to Jeffreys, the Earth is simply too rigid to allow for any significant movement of the crust.

The opponents of continental drift offered their own suggestions for how, in some cases, the flora and fauna of far-separated continents could manifest remarkable similarities. For example, the continents in question may at one time have been connected by slender land bridges, similar to that which used to connect Alaska and Siberia across the Bering Strait. One opponent – Chester Longwell, mentioned earlier – even made the desperate suggestion that:[7]

> If the fit between South America and Africa is not genetic, surely it is a device of Satan for our frustration.

In short, there were two clashing opinions – two competing paradigms. Each paradigm faced questions which it could not answer in any fully satisfactory way – questions of coincidence, and questions of mechanism. In such a case, the opinions adopted by leading scientists depended at least in part on their background philosophies of life, rather than on the intrinsic significance of any one piece of evidence. Science historian Naomi Oreskes points to a couple of factors that were particularly significant for at least some leading American geologists:[8]

> For Americans, right scientific method was empirical, inductive, and required weighing observational evidence in light of alternative explanatory possibilities. Good theory was also modest, holding close to the objects of study... Good science was anti-authoritarian, like democracy. Good science was pluralistic, like a free society. If good science provided an exemplar for good government, then bad science threatened it. To American eyes Wegener's work was bad science: It put the theory first and then sought evidence for it. It settled too quickly on a single interpretive framework. It was too large, too unifying, too ambitious. In short, it was seen as autocratic...
>
> Americans [also] rejected continental drift because of the [principle] of uniformitarianism. By the early twentieth century, the methodological principle of using the present to interpret the past was deeply entrenched in the practice of historical geology. Many believed this the only way to interpret the past, that uniformitarianism made geology a science, for without it what proof was there that God hadn't made the Earth in seven days, fossils and all?... but according to drift theory, continents in tropical latitudes did not necessarily have tropical faunas, because the reconfiguration of continents and oceans might change matters altogether. Wegener's theory raised the spectre that the present was not the key to the past—that it was just a moment in Earth history, no more or less characteristic than any other. This was not an idea Americans were willing to accept.

Changing Minds on Moving Continents

Deep Learning pioneer Geoffrey Hinton provides one additional account of the entrenched resistance to the idea of continental drift. He explains about the experience of his father, who was an entomologist (a specialist in the study of insects):[9]

> My father was an entomologist who believed in continental drift. In the early '50 s, that was regarded as nonsense. It was in the mid-50 s that it came back. Someone had thought of it 30 or 40 years earlier named Alfred Wegener, and he never got to see it come back. It was based on some very naive ideas, like the way Africa sort of fit into South America, and geologists just pooh-poohed it. *They called it complete rubbish, sheer fantasy.*
>
> I remember a very interesting debate that my father was involved in, where there was a water beetle that can't travel very far and can't fly. You have these

in the north coast of Australia, and in millions of years, they haven't been able to travel from one stream to another. And it came up that in the north coast of New Guinea, you have the same water beetle, with slight variations. The only way that could have happened was if New Guinea came off Australia and turned around, that the north coast of New Guinea used to be attached to the coast of Australia. It was very interesting seeing the reaction of the geologists to this argument, which was that 'beetles can't move continents.' *They refused to look at the evidence.*

The above descriptions may lead to the conclusion that an unresolvable impasse had been reached, with people in different paradigms unwilling even to look at the evidence they couldn't explain. Indeed, an impasse persisted for several decades. Then, thankfully, good science prevailed. Despite the stubbornness of a number of individual scientists, the science community as a whole remained open to the possibilities of significant new evidence; and significant new evidence emerged.

First, in the 1950s, geologists started paying more attention to the emerging field of paleomagnetism.[10] That field looks at the orientation of magnetic material in rocks or sediment. This orientation was seen to vary in prehistoric rocks from that of more recent rocks, with interesting patterns of variation that became apparent due to improved measurement techniques. Scientists were led to conclude: either the earth's magnetic poles were in a different location when these rocks were formed, or the rocks may have moved over large parts of the earth in the intervening aeons of time. The more closely geologists looked at this data, the more support they found for the principle of continental drift. For example, rock samples from India strongly suggested that India had previously lain south of the equator (whereas nowadays it is entirely north of that line).

Second, examination of the trenches on the ocean floor, along with deep thermal vents and submarine volcanoes, provided further evidence of significant subterranean fluid activity. This helped establish the concept that continental plates were propelled apart by the sea-floor spreading. What decided the matter, for many scientists, was the result of a particular test that was proposed. Oreskes takes up the story:

> Meanwhile, geophysicists had demonstrated that the earth's magnetic field has repeatedly and frequently reversed its polarity. Magnetic reversals plus sea-floor spreading added up to a testable hypothesis…: If the sea floor spreads while the Earth's magnetic field reverses, then the basalts forming the ocean floor will record these events in the form of a series of parallel 'stripes' of normal and reversely magnetized rocks.

Since World War II, the United States Office of Naval Research had been supporting sea-floor studies for military purposes, and large volumes of magnetic data had been collected. American and British scientists examined the data, and by 1966 the… hypothesis had been confirmed. In 1967–68, the evidence of drifting continents and the spreading sea-floor was unified into a global framework.

In the end, scientific consensus changed relatively quickly, as more and more data was linked with refined, more sophisticated models of sea floor spreading and the resulting continental drift.

In parallel, the strong philosophical positions which had previously predisposed some scientists to oppose the theory of continental drift – positions such as a preference for "modest" theories, and the preference for uniformitarianism over any sort of catastrophism – had lost their vigour. These philosophies were recognised as being, perhaps, useful general guides, but lacking the universality to strike down theories which had strong explanatory and predictive powers of their own.

Washing Hands

The principles that applied in the case of continental drift also applied in the case of hand disinfection in hospitals. If Alfred Wegener was the sad victim in the first case – dying in obscurity in Greenland, in 1930, long before the merit of his hypothesis was widely recognised – then Ignaz Semmelweis was similarly the victim in the second case.

Semmelweis had gathered experimental data in favour of improved sanitation in hospitals, but his theories found little respect at the time. He became severely depressed, and was confined to a mental institution where he was beaten by guards and confined in a straitjacket. He died within two weeks of entering the institute, aged only 47.[11]

Some twenty years earlier, in 1846, the young Semmelweis had been appointed to an important medical assistance role in the maternity department of Vienna's General Hospital. The hospital had two maternity clinics, and townspeople already knew that the mortality rate at one of these clinics (10% and upwards) was considerably higher than at the second (4%). Large numbers of women in the first clinic were dying from puerperal fever (childbed fever) after giving birth. Semmelweis put a lot of effort into trying to understand this variance. He finally observed that medical students working at the first clinic frequently also performed autopsies on cadavers, before visiting the maternity ward and examining women there; no such students worked in the second ward. It was a keen piece of empirical observation.

Based on his observations, Semmelweis surmised that some kind of microscopic material from dead bodies, carried on the hands of trainee doctors, was the cause of the high mortality rate in the first clinic. He introduced a system of rigorous hand-washing, using chlorinated lime. This process removed the odour of dead bodies from doctors' hands, in a way not accomplished by conventional washing with soap and water. The death rate plummeted, reaching zero within one year.

From our modern standpoint, we are inclined to say "Of course!" We find ourselves astonished at the prior lack of hand-washing. However, all this took place several decades before Louis Pasteur popularized the germ theory of disease. At that time, it was commonly thought that diseases were spread by "bad air" (miasma). Indeed, lacking any awareness of germs, the medical orthodoxy of the time resisted the advice of Semmelweis that rigorous hand-washing be introduced more widely.

In an echo of criticisms that were to be applied one hundred years later to Alfred Wegener, the ideas of Semmelweis were seen as being too all-encompassing: too far-reaching, and too disruptive. Semmelweis claimed that one single cause – poor cleanliness – was responsible for a large proportion of hospital illnesses. That flew in the face of prevailing medical doctrine, which held that each individual case of illness had its own unique causes, and therefore needed its own tailored investigation and treatment. Blaming everything on poor hygiene was too singular an idea.

The practice of painstaking hand-washing was also something that, it seems, offended at least some doctors, who were affronted by the idea that their normal gentlemanly levels of personal hygiene might somehow be substandard. They could not accept that they, personally, were responsible for the deaths of the patients they examined.

Semmelweis lost his position at Vienna's General Hospital in 1848, the year of many revolutions throughout Europe. The head of the department was politically conservative, and increasingly distrusted Semmelweis, some of whose brothers were actively involved in the movement for Hungarian independence from Austria. This political difference exacerbated an already fraught personality conflict. Semmelweis left the hospital and was replaced in his role by Carl Braun. Remarkably, Braun undid much of the progress in the clinic. Braun later published a textbook that listed thirty different causes of childbed fever. The mechanism identified by Semmelweis, poisoning from microscopic material from corpses, featured as number 28 on the list,[12] with little prominence. Maternal death rates rose again in the clinic, as focus on proper hygiene was replaced by a predilection for improved ventilation systems – a predilection that fitted the prevailing miasma ("bad air") paradigm for the cause of many diseases.

Therefore, even at the hospital where the breakthrough insight had occurred, the heavy weight of orthodox tradition resulted in numerous women subsequently dying needless deaths. Similar dismal patterns were followed throughout Europe, until such time as independent evidence in favour of the germ theory of disease had accumulated, through the work of (among others) John Snow, Joseph Lister, and Louis Pasteur. By 1880s, thorough antiseptic washing had become standard practice, and the miasma paradigm had been overturned by the germ theory one.

Not for the first time – and not for the last – established practice within the medical profession, therefore, fell far short of the founding principle of the occupation: *first, do no harm*. Faulty thinking by doctors led to poor hygiene and therefore to an avalanche of unnecessary harm. This departure from the Hippocratic Oath was in part due to lack of knowledge (lack of the germ theory of disease) but also due to the overhang of prior habits and prior styles of thinking.

In our view, the accepting-aging paradigm fits the same pattern. It persists in part due to lack of knowledge (the progress made by rejuvenation biotechnology), but also due to the overhang of prior habits and prior thinking styles. Those who are immersed in that paradigm tend to see things differently, of course.

Medical Paradigm Shifts, Resisted

Ignaz Semmelweis is often identified as a key pioneer of the broader principle of "evidence-based medicine". He tested his hypotheses about the causes of childbed fever by varying the practice of medical staff and observing the subsequent changes in mortality. His observations had previously ruled out a number of potential causes of the differences in mortality in the two clinics – different socio-economic status, different physical positions adopted by the mothers while giving birth, and so on. When the new antiseptic hand-washing routine was introduced, the results were dramatic.

As we've seen, however, this evidence failed to make sense from inside the competing paradigm which viewed "bad air" as the more likely cause of diseases. Supporters of that paradigm reinterpreted the changed mortality rates as likely having different causes – such as improved ventilation. Unfortunately, no rigorous tests were carried out to distinguish between these different theories. The principles that we nowadays expect to apply to trials of medical efficacy were not yet understood at that time, despite the insight of Semmelweis. These principles include:

- Control: patients receiving a new treatment are compared with a "control" set who don't receive that treatment (they may receive a placebo instead), but who are otherwise as alike as possible to the first group
- Randomisation: the assignment of patients between the two groups, control and treated, takes place randomly, to prevent biases (conscious or unconscious) in the selection that would undermine the result
- Statistical significance: the design of tests to avoid being misled by the chance deviations that naturally occur from time to time; in particular, tests with small sample sizes have little value
- Reproducibility: repetition of trials, with different sets of clinical practitioners involved each time; if the same results arise, this gives a greater indication of the underlying reliability of a proposed treatment.

Indeed, the term "evidence-based medicine" is only a few decades old. The first academic paper published about it was in 1992.[13] The term was introduced in distinction to the prevailing practice of "clinical judgement", which refers to doctors taking decisions on potential treatments based upon their own hunches and intuitions – hunches and intuitions that have in turn been schooled by the long experiences of individual doctors. An alternative term for "clinical judgement" commonly in use was "the art of medicine".

The drawbacks of reliance on "clinical judgement" were forcibly underlined in a 1972 book *Effectiveness and Efficiency: Random Reflections on Health Services* by Scottish doctor Archie Cochrane.[14] Cochrane was passionately critical about much of the thinking and practice of his fellow medical professionals. He pointed out that:

- A significant part of the earlier improvement in public health was due to improvement in environmental factors, such as hygiene, rather than medical treatments in their own right
- Doctors are under great pressure from their patients to provide them a prescription or some other treatment, and may well do so, even though there's no clinical evidence for the effectiveness of that treatment
- The fact that some patients recover after being treated by a particular course of treatment is no proof of the effectiveness of that treatment; the recovery might instead be caused by other factors (including the body's tendency to get better of its own accord, in time)

- The fact that patients believe a course of treatment has done them good is, again, no proof of the effectiveness of that treatment.

Cochrane noted that culture in general, at the time he wrote his book, was more likely to be impressed by "opinion" than by "experiment":[15]

> There still seems to be considerable misunderstanding amongst the general public and some medical people about the relative value of opinion, observation, and experiment, in testing hypotheses.
>
> Two of the most striking changes in word usage in the last twenty years are the upgrading of 'opinion' in comparison with other types of evidence, and the downgrading of the word 'experiment'. The upgrading of 'opinion' has doubtless many causes, but one of the most potent is, I am sure, the television interviewer and producer. They want everything to be brief, dramatic, and black and white. Any discussion of evidence is written off as lengthy, dull, and grey. I have seldom heard a television interviewer ask anyone what his evidence was for some particular statement. Fortunately it does not usually matter; the interviewers only want to amuse (hence the interest in pop singers' views on theology), but when they deal with medical matters it can be important.
>
> The fate of 'experiment' is very different… It has been taken over by journalists and debased… and is now being used in its archaic sense of 'action of trying anything', hence the endless references to 'experimental' theatres, art, architecture, and schools.

Cochrane had plenty of good things to say about medical practice. He described some positive examples which could serve as templates for future investigations – for example, the development of effective treatments for tuberculosis, which involved wide usage of randomised control trials in the years after the Second World War. He praised doctors for being far ahead of other professionals, such as judges and headmasters, in organising experimental controlled trials of different "therapeutic" or "deterrent" treatments. However, as Cochrane pointed out, medical history is full of examples where strongly prevailing opinion was eventually demonstrated, via careful experiments, to be incorrect:

- Tonsillectomy, especially for children, was once thought to be a near-panacea, and was practised widely, but following a critical review of the evidence in 1969 (in an article entitled "Ritualistic surgery – circumcision and tonsillectomy"), it is now undertaken much less often

- The gold-based compound sanocrysin became popular in America in the 1920s as a treatment for tuberculosis; one doctor published in 1931 the results of a trial with 46 patients in which he declared that the drug was "outstanding". However, that trial contained no controls; all 46 patients received the drug. In the same year, other doctors, from Detroit, trialled the drug on a randomly chosen subset of 12 out of 24 tuberculosis sufferers. The other patient in each pair instead received an injection that just contained sterile water, unbeknown to them. The result this time was decisive: the control patients were the ones more likely to survive. Sanocrysin, previously acclaimed as a wonder drug, was shown to be nothing of the sort
- Enforced bed rest was another treatment for tuberculosis that was long popularised, until tests in the 1940s and 1950s showed that such a regime was actually harmful rather than beneficial: patients that lay supine had extra complications from their coughs. Sanatoria around the world were closed in the wake of this research.

In parallel, Cochrane demonstrated instances where supposed clinical expertise was nothing like as infallible as its practitioners claimed. Druin Burch recounts the following episode in his 2009 book *Taking the Medicine: A Short History of Medicine's Beautiful Idea, and our Difficulty Swallowing It:*[16]

> Electrocardiograms (ECGs) are recordings of the heart's electrical activity… Cardiologists claim skills in reading them that are beyond the measure of other doctors. Cochrane took randomly selected ECGs and sent copies to four different senior cardiologists, asking them what the tracings showed. He compared their opinions and found that these experts agreed only 3 per cent of the time. Their confidence in being able to look at the tracings and see the 'truth' did not seem justified. At least ninety-seven times out of a hundred, someone was getting something wrong.
>
> When Cochrane performed a similar test with professors of dentistry, asking them to evaluate the same mouths, he found that there was only a single thing that their diagnostic skills consistently agreed on: the number of teeth.

After his death in 1988, Cochrane's surname was incorporated in 1993 in the name of the newly founded Cochrane Collaboration. The collaboration describes its work as follows:[17]

> Cochrane exists so that healthcare decisions get better.
>
> During the past 20 years, Cochrane has helped to transform the way health decisions are made.

We gather and summarize the best evidence from research to help you make informed choices about treatment...

Cochrane is for anyone who is interested in using high-quality information to make health decisions. Whether you are a doctor or nurse, patient or carer, researcher or funder, Cochrane evidence provides a powerful tool to enhance your healthcare knowledge and decision making.

Cochrane contributors – 37,000 from more than 130 countries – work together to produce credible, accessible health information that is free from commercial sponsorship and other conflicts of interest.

The Cochrane Collaboration, in fulfilling the vision of evidence-based medicine that was trail-blazed by Archie Cochrane and others, is nowadays recognised as performing an extremely important task. By 2009, *Cochrane Reviews* were being downloaded from their website at the rate of one every three seconds.[18] Among the currently most popular downloads are reviews on the evidence in subjects such as:[19]

* Acupuncture for tension-type headache
* Midwife-led continuity models versus other models of care for child-bearing women
* Interventions for preventing falls in older people living in the community
* Vaccines to prevent influenza in healthy adults.

These are all areas where intuitive "clinical judgement" is very usefully supplemented by a careful survey of the experimental evidence – evidence which often confounds expert expectations.

Without being aware of the history, it would be difficult to imagine how much hostility the concept of evidence-based medicine engendered before it became more widely accepted. The original criticism of clinical judgement was widely resisted:

* Senior medical professionals feared that their hard-won tacit knowledge would become under-valued by the movement towards black-and-white evidence-based medicine
* These same professionals often insisted that patients needed to be treated as individuals, rather than being forcibly fit into one of a small number of stereotypes as featured in new medical textbooks.

Bloodletting

Let's look at one final telling example. Bloodletting – the removal of blood from a patient's body, often by use of leeches – was widely advocated as a medical treatment for more than two thousand years. It was recommended for a huge string of medical conditions, including acne, asthma, diabetes, gout, herpes, pneumonia, scurvy, smallpox, and tuberculosis. Early prominent supporters included Hippocrates of Kos (460-370 BC) and Galen of Pergamum (129-200 AD). The practice had its occasional prominent critics over the centuries, including William Harvey, the person who had in the 1620s discovered the path of the circulation of blood around the body, but it continued in wide use. DP Thomas, writing in 2014 in the Journal of the *Royal College of Physicians Edinburgh*, notes that:[20]

> The fervour with which physicians in earlier times carried out bloodletting seems extraordinary today. Guy Patin (1601–1672), Dean of the Paris Medical Faculty, bled his wife 12 times for a 'fluxion' of the chest, his son 20 times for a continuing fever, and himself seven times for a 'cold in the head'. Charles II (1630–1685) was bled following a stroke, and General George Washington (1732–1799), suffering from a severe throat infection, was bled four times in a matter of a few hours. The amount of blood taken from him has been variously estimated at between five and nine pints. Strong man though he was, even his constitution could not withstand the misguided efforts of his physicians, and it seems likely such treatment hastened his end.

Thomas goes on to mention the case of Benjamin Rush:

> Benjamin Rush (1746–1813), a distinguished American physician and signer of the Declaration of Independence, was convinced that bleeding his patients was the best treatment... During the yellow fever epidemic in Philadelphia in 1793 Rush bled and purged his patients...
> Rush's approach is a salutary reminder of the dangers of sincerely held beliefs in the value of traditional methods, and highlights the need for a critical, evidence-based assessment of all forms of treatment.

Systematic evidence on the effect of bloodletting started to be gathered in the nineteenth century. Frenchman Pierre Charles Alexandre Louis analysed data in 1828 from 77 patients suffering pneumonia, showing that bloodletting had, at best, little effect on the prospects for recovery. However, many practising physicians pushed his results aside, preferring to rely on what they *thought* their own personal experience confirmed, and to trust the weight of venerable tradition running all the way back to Hippocrates and Galen.

In the second half of the nineteenth century, John Hughes Bennett at Edinburgh University reviewed additional data on survival rates in hospitals in American and in Britain. He pointed out that, for example, over an 18-year period at Edinburgh Royal Infirmary, out of 105 standard cases of pneumonia that he himself treated, without any bloodletting, not one patient died. In contrast, at least one third of the patients who did receive bloodletting, under treatment from other physicians at the hospital, subsequently died. But despite this data, Hughes Bennett faced fierce criticism from within his own profession. DP Thomas comments:

> From today's perspective, perhaps the most surprising aspect of the pioneering work of Louis and Hughes Bennett was how slow the medical profession was to accept their strong evidence, especially in relation to the treatment of pneumonia. Hughes Bennett was attempting to introduce a more scientific approach to identifying and treating disease, involving both laboratory observations and statistical analysis of results. However, this approach came into conflict with that of more traditional clinicians who continued to rely on their own experience, based solely on clinical observation. Despite growing scepticism of the treatment, the controversy about bloodletting continued throughout the latter half of the nineteenth century, and indeed well into the twentieth.

Writing in the *British Columbia Medical Journal* in 2010, Gerry Greenstone reflects on the question as to why bloodletting continued for so long, well into the middle of the twentieth century:[21]

> We may wonder why the practice of bloodletting persisted for so long, especially when discoveries by Vesalius and Harvey in the 16th and 17th centuries exposed the significant errors of Galenic anatomy and physiology. However, as IH Kerridge and M Lowe have stated, "that bloodletting survived for so long is not an intellectual anomaly—it resulted from the dynamic interaction of social, economic, and intellectual pressures, a process that continues to determine medical practice."
>
> With our present understanding of pathophysiology we might be tempted to laugh at such methods of therapy. But what will physicians think of our current medical practice 100 years from now? They may be astonished at our overuse of antibiotics, our tendency to polypharmacy, and the bluntness of treatments like radiation and chemotherapy.

Never mind "100 years from now". In our view, it's likely that within 10–20 years, physicians will be looking back at present-day practice with astonishment that the phenomenon of aging received so little attention, and that rejuvenation biotechnology was such a minority interest.

But as we have already seen, paradigms have deep effect. The phrase from IH. Kerridge and M Lowe quoted above puts the same sentiment in other words: medical practice arises "from the dynamic interaction of social, economic, and intellectual pressures".

Everyone can be wrong, not just the experts, of course. To ironically close this chapter, let us remember the striking reflection of the American beauty queen Heather Whitestone, who was Miss Alabama 1994 and then Miss America 1995. When questioned during the beauty pageant whether she wanted to live forever, she answered:[22]

> I would not live forever, because we should not live forever, because if we were supposed to live forever, then we would live forever, but we cannot live forever, which is why I would not live forever.

Notes

1. http://mathworld.wolfram.com/Rabbit-DuckIllusion.html
2. http://www.moillusions.com/vase-face-optical-illusion/
3. http://well.blogs.nytimes.com/2008/04/28/the-truth-about-the-spinning-dancer/?_r=0
4. http://www.amazon.com/Alfred-Wegener-Creator-Continetal-Science/dp/0816061742/
5. https://www.smithsonianmag.com/science-nature/when-continental-drift-was-considered-pseudoscience-90353214/
6. https://www.e-education.psu.edu/earth520/content/l2_p12.html
7. http://folk.ntnu.no/krill/krilldrift.pdf
8. http://www.mantleplumes.org/WebDocuments/Oreskes2002.pdf
9. https://www.macleans.ca/society/science/the-meaning-of-alphago-the-ai-program-that-beat-a-go-champ/
10. http://geologylearn.blogspot.com/2016/02/paleomagnetism-and-proof-of-continental.html
11. https://embryo.asu.edu/pages/ignaz-philipp-semmelweis-1818-1865
12. https://en.wikipedia.org/wiki/Carl_Braun_(obstetrician)#Views_on_puerperal_fever
13. http://jama.jamanetwork.com/article.aspx?articleid=400956

14. http://www.amazon.com/Effectiveness-Efficiency-Random-Reflections-Services/dp/185315394X/
15. https://www.nuffieldtrust.org.uk/files/2017-01/effectiveness-and-efficiency-web-final.pdf
16. http://www.amazon.com/Taking-Medicine-Medicines-Difficulty-Swallowing/dp/1845951506/
17. http://www.cochrane.org/about-us
18. https://community.cochrane.org/sites/default/files/uploads/inline-files/Cochrane%20funding%20brochure_A5_print_12May16.pdf
19. http://www.cochrane.org/evidence
20. https://www.rcpe.ac.uk/sites/default/files/thomas_0.pdf
21. http://www.bcmj.org/premise/history-bloodletting
22. https://www.mtechnologies.com/n1fn/bcramps.htm

8

Plan B: Cryopreservation

Being cryopreserved after death is the second-worst thing that could happen.
The worst thing is dying without being cryopreserved.
Ben Best, 2005

If cryonics were a scam it would have far better marketing and be far more
popular.
Eliezer Yudkowsky, 2009

Cryonics is an experiment. Do you want to join the control group, or the
experimental group?
Ralph Merkle, 2017

We estimate that the first biotech treatments for human rejuvenation will be commercialized in the 2020s, followed by nanotech treatments in 2030, and then full control and reversal of aging by 2045. Until then, unfortunately, people will continue to die. For most people who have ever lived, rejuveneering is coming too late. It's coming too late for them, either because they've already grown old and died, in the present century or in one of the many preceding ones, or because, although they're still alive, they're likely to die before effective rejuvenation therapies become widely available. Either way, they belong to the BR era – the Before-Rejuvenation era.

However, within the broad rejuveneering community, some researchers dare to suggest there might be hopes for the regeneration of people from the BR era. Collectively, these ideas form a set of radical alternatives and complements to the "Plan A" methods we've covered in previous chapters in this book.

A Bridge to Eternity

As we have discussed before, an indefinite lifespan will be possible in a few decades, but what can we do until then? The sad truth is that people will continue to die for years to come, and the best way we know today to preserve ourselves relatively well is through cryopreservation. We could say that cryopreservation is a "Plan B" of indefinite human lifespan until Plan A arrives.

The modern era of human cryopreservation, or simply cryonics, began in 1962 when the American physicist Robert Ettinger published *The Prospect of Immortality*, where he considered that freezing (actually: cryopreserving) patients in anticipation of the future arrival of much more advanced medical technologies to cure current diseases, including aging.[1] Although cryopreservation of a human being may seem fatal, Ettinger argued that what appears fatal today may be reversible in the future. The same argument applies to the process of death itself, i.e., the early stages of clinical death may be reversible in the future. Combining these ideas, Ettinger suggests that freezing recently deceased people may be a way to save lives. Based on these ideas, Ettinger and four other colleagues founded the *Cryonics Institute* in 1976 in Detroit, Michigan. Their first patient was Ettinger's mother, who was cryopreserved in 1977. Her body is kept frozen at the boiling temperature of liquid nitrogen (-196 °C).

Meanwhile, in California, Fred and Linda Chamberlain founded another cryopreservation institution in 1972 under the name *Alcor Life Extension Foundation* (originally and until 1977 called *Alcor Society for Solid State Hypothermia*). Their first patient, in 1976, was Fred Chamberlain's father, who underwent a neuropreservation in which only the head was cryopreserved. Alcor eventually moved in 1993 to Scottsdale, Arizona, far from seismic California, and its President for many years was the English philosopher and futurist Max More, who explains:[2]

> We see it as an extension of emergency medicine... We're just taking over when today's medicine gives up on a patient. Think of it this way: 50 years ago if you were walking along the street and someone keeled over in front of you and stopped breathing you would have checked them out and said they were dead and disposed of them. Today we don't do that, instead we do CPR and all kinds of things. People we thought were dead 50 years ago we now know were not. Cryonics is the same thing, we just have to stop them from getting worse and let a more advanced technology in the future fix that problem.

Several patients decide just to preserve their head. Some do so for economic reasons; others believe that human identity and memory are stored in the brain and therefore it is not necessary to cryopreserve the entire body, which can also be reconstructed using different technologies.

As for *Cryonics Institute*, it only makes total cryopreservations, while *Alcor* makes both neuropreservations and complete cryopreservations. To date, *Cryonics Institute* has over 200 cryopreserved patients[3] and over a thousand members, while *Alcor* has a similar number of patients[4] (of which about three-quarters are neuropatients) and members. Each month new patients and members join the two main cryopreservation centers in the United States. Both institutions also keep many frozen samples of DNA, tissues, pets, and other animals under cryopreservation. *Cryonics Institute* charges between $28,000 and $35,000 (not including the high costs of *SST*: *Standby/Stabilization/Transport*) for full-body cryopreservations. Alcor charges $80,000 for neuropreservations and $200,000 for full-body cryopreservations (including the high costs of *SST*).[5]

Given that the number of patients and members is still relatively small, *Cryonics Institute* and *Alcor* were virtually the only two cryopreservation organizations in the world until 2005, when KrioRus was founded outside Moscow. Today there are also small groups in Argentina, Australia, Canada, China, Germany, and the states of California, Florida, and Oregon in the United States that plan to create or have already created new storage facilities for human cryopreservation. Shandong Yinfeng Life Science Research Institute was founded in 2015 and already has a dozen patients in China. Southern Cryonics in Australia and the European Biostasis Foundation in Switzerland have opened cryonic centers in 2022, with no patients as of this writing.

How Does Cryonics Work?

So far no one has revived after being cryopreserved, but it is also because we still do not know how to cure the conditions that caused the terminal illness that the patient suffered his time. However, thanks to exponential technological advances, it is very likely that we will be able to revive patients in the coming decades. The American futurist Ray Kurzweil talks about the 2040s for the first reanimation of cryopreserved patients, starting with the last to have undergone this technique, who will have been cryopreserved with better technologies, and ending with the first patients.[6]

The proof of concept is that cryopreservation has already been done with different living cells, tissues, and small organisms. Tiny tardigrade water bears are microscopic multicellular organisms that can survive if most of their internal water is replaced by trehalose sugar, which prevents crystallization of cell membranes. Several vertebrates also tolerate freezing, and some organisms survive the winter by solid freezing and ceasing their vital functions. Some species of frogs, turtles, salamanders, snakes, and lizards can survive freezing and recover completely after wintering in cold climates. Some species of bacteria, fungi, plants, fish, insects, and amphibians living near the poles have developed cryoprotectants that allow them to survive in freezing conditions.

British scientist James Lovelock, known for proposing the Gaia hypothesis about life on Earth, was perhaps the first person to attempt to freeze and reanimate animals. In 1955, Lovelock froze some rats at 0 °C and then successfully resuscitated them using microwave diathermy. Recently, DARPA, The Defense Advanced Research Projects Agency, has begun funding research on suspended animation, essentially "turning off" the heart and brain so that appropriate treatment can later be offered to certain patients, something that can be considered a step in the cryopreservation of human beings.

Egg, sperm and even embryos are cryopreserved today to be resuscitated in the future. Frozen ova and sperm have been used for animal reproduction, and numerous human embryos have been cryopreserved and subsequently developed without congenital or any other problems. Also, blood, umbilical cords, bone marrow, plant seeds, and various tissue samples are now routinely frozen and thawed. One of the great recent successes of cryonics was the birth in 2017 of an embryo cryopreserved for almost 25 years.

We believe that people cryopreserved today can be revived in the future through the use of advanced techniques. There is a growing body of scientific literature that supports the viability of cryonics. Some prestigious scientists signed an open letter supporting cryonics, including Aubrey de Grey and American scientist Marvin Minsky, considered one of the "fathers" of artificial intelligence, who was cryopreserved upon his death in 2016:[7]

Cryonics is a legitimate science-based endeavor that seeks to preserve human beings, especially the human brain, by the best technology available. Future technologies for resuscitation can be envisioned that involve molecular repair by nanomedicine, highly advanced computation, detailed control of cell growth, and tissue regeneration.

With a view toward these developments, there is a credible possibility that cryonics performed under the best conditions achievable today can preserve sufficient neurological information to permit eventual restoration of a person to full health.

The rights of people who choose cryonics are important, and should be respected.

In 2015, a group of scientists from the universities of Liverpool, Cambridge and Oxford established a cryonics research network in the United Kingdom for the encouragement and promotion of cryonics research and its applications, including human cryopreservation.[8] Thanks to these advances, more and more people around the world are beginning to realize that human cryopreservation is possible, especially as proof of concept is now available.

Cryonics in Russia: A Visit to KrioRus

Those familiar with cryonics will have heard about the two major cryopreservation facilities in the United States: *Cryonics Institute* near Detroit, Michigan, and *Alcor* in Scottsdale, Arizona. But not so many know that a cryonics organization was founded outside Moscow in 2005 under the leadership of Russian futurist Daniel Medvedev.

During a meeting with Medvedev in 2015, we were able to visit the facilities of KrioRus of Sergiyev Posad, a beautiful and ancient city about 70 km northeast of Moscow. Sergiyev Posad is known for being a religious and tourist destination where you will find one of the largest Russian monasteries, the Trinity Lavra of Saint Sergius monastery, founded by Saint Sergius of Radonezh in the fourteenth century. Sergiyev Posad seems a very appropriate place for cryopreservation as it was for the traditional resting place of Russian saints and monarchs. KrioRus has grown rapidly and is considering the expansion of its facilities or moving to another location also near Moscow, where they could also establish a hospice and ancillary facilities for terminal patients, in addition to a cryopreservation facility with more research capacity.

The growth of KrioRus has been spectacular compared to Alcor and Cryonics Institute. In just over a decade, KrioRus managed to cryopreserve over half a hundred people and dozens of pets, including many dogs, cats, and birds, as well as a Chinchilla rodent. KrioRus' first patient was Lidiya Fedorenko, in 2005, who was originally cryopreserved on dry ice for a few months until the first container or "cryostat" was ready. Medvedev's grandmother is another patient currently under neuropreservation. Like the Cryonics Institute, KrioRus uses cryostats, which are large containers made of fiberglass/resin filled with liquid nitrogen instead of the more expensive individual Dewar flasks used by Alcor. All patients, pets, and tissues cryopreserved in KrioRus are stored in two large cryostats specially designed by KrioRus, which has gained enough experience to build new cryostats for its new facilities, in addition to plans to install a new center in Switzerland.

KrioRus charges 12,000 euros for neuropreservations and 36,000 euros for full-body cryopreservations, not including stand-by/stabilization/transport costs, which vary greatly according to the patient's place of origin. Cryopreservation of animals and tissues is cheaper, depending on size and other special conditions. In the last decade, KrioRus has managed to attract patients not only from Russia, but also from other European countries, such as Italy, the Netherlands, and Switzerland, and from much further afield, such as Australia, Japan, and the United States. As in the case of Alcor, more than half of cases are neuropatients. The relatively rapid growth of KrioRus indicates that effective and affordable service can help popularize cryopreservation.

Again, we insist that life arose for the sake of life, not for the sake of death. We expect a cure for aging by midcentury, but for that to happen, it is essential to declare war on aging. Meanwhile, cryonics is Plan B. Proofs of concept already exist that both indefinite lifespans are possible (Plan A) and cryonics is also possible (Plan B). Now we need more scientific advances to solve technical problems because we know that it is possible, and the sooner we do it, the better for humanity. Every life that is lost is a tragedy, no doubt a personal tragedy, as well as a loss for society as a whole, but we can stop it. As the death of death arrives, we move towards an indefinite human lifespan: Long live cryopreservation!

An Ambulance to the Future

One of the most important innovations in medical history can be said to be the creation of the ambulance. If someone is injured, or suffers a medical emergency, the timely arrival of an ambulance can make all the difference between life and death. Such a person was the victim of "being in the wrong place"; they needed medical help, but they weren't in a location where medical help was immediately available. However, an ambulance meant they could be transported to a facility with the resources to treat them: equipment, medicines, and trained healthcare professionals.

There may be scope for observers to quibble about the costs of any given ambulance service. Some critics might say that an ambulance service could be provided at less cost, whilst still being able to meet the majority of the demands placed upon it. However, it's rare to find someone who complains about the very idea of an ambulance service. You don't hear people saying that if someone experiences a medical emergency away from a hospital, that's simply too bad – they should stoically accept their fate. Nor is it said that any family members who request an ambulance for their injured parent, child, or sibling are being selfish or immature. Instead, society embraces the idea that it is natural to demand speedy, safe transport away from the initial danger zone into a place where the medical emergency can be properly dealt with. That way, the injured person has a chance to be treated, and to live on, potentially for many decades to come.

But consider our attitude towards someone who suffers a medical emergency at "the wrong time". They happen to have a disease which is about to kill them, but which medical science is likely to be able to cure in, say, thirty years' time. What should we think about the provision of a possible "ambulance to the future" for such a person? Suppose, for the sake of argument, there was a chance of at least 5% that such an "ambulance" might work. To be precise, the mechanism under consideration is the low-temperature cryonic preservation of the person, in such a way that they enter something akin to deep coma in which all their normally physiological processing is suspended. Should we embrace the possibility of such a rescue vehicle? Or should we instead urge the victim of this medical crisis to avoid thinking about any such possibility? Should we tell them, in other words, to stoically accept their fate (their impending death)? And if any of their family members, wishing to be able to converse and interact with the dying person in the future, request the provision of this kind of ambulance service, should we rebuke them for being selfish or immature?

Of course, the analogy is far from perfect. With an ambulance that transports a patient through space, to a hospital, there are plenty of previous examples of the journey succeeding. But with cryonics, the journey of a human patient through decades of low-temperature bodily suspension has never yet completed. We can see photographs of the storage cylinders inside which cryonics patients are preserved. But there is no guarantee that medical science will ever progress to the point where these patients can be successfully reanimated.

The arguments against the idea of cryonics mirror the arguments against rejuveneering. Some critics say cryonics *cannot succeed*: the technical challenges of wakening someone from such a low temperature state are unfathomably difficult. The process of lowering the body to ultra-low temperatures may have irreparably damaged it, despite the careful use of anti-freeze, cryoprotectants, and other sophisticated chemicals. After all, these chemicals are toxins in their own right, and the process of cooling large organs may introduce fractures. Other critics say that cryonics *should not even be considered*, because it is morally wrong. They allege that it's a misuse of valuable resources, a wicked delusion, a financial scam, or worse.

Our response to these criticisms is – like our response to the corresponding criticisms of rejuveneering – to profoundly disagree. In both cases, we view the majority of the criticisms to be ill-informed, or to be motivated by faulty reasoning and other (often repressed) ulterior drives. In both cases – rejuveneering and cryonics – we accept that the engineering task will be hard. But we see no reason in either case why the task will be impossible. In time, high quality solutions can be created. In both cases, we see a series of precursors already in place, which point the way towards an eventual comprehensive engineering solution.

One precursor for cryonics is the field known as therapeutic hypothermia. In 1999, trainee doctor Anna Bågenholm was skiing off-piste in a steep descent in a remote region of northern Norway, when she fell into a frozen mountain stream. By the time a rescue helicopter arrived, she had been in freezing water for 80 minutes, and her blood circulation had been stopped for 40 minutes, as reported in a subsequent article in the Lancet, "Resuscitation from accidental hypothermia of 13·7°C with circulatory arrest".[9] Writing in the Guardian, in an article entitle "Between life and death – the power of therapeutic hypothermia", David Cox provides further details:[10]

> By the time Bågenholm was brought to the University Hospital of North Norway in Tromso, her heart had stopped for well over two hours. Her core temperature had plunged to 13.7°C. She was in every sense clinically dead.
>
> However, in Norway, there has been an old saying for the past three decades that you're never dead until you're warm and dead. Mads Gilbert is the head of emergency medicine at the hospital and, from experience, he knew that there was a slim chance the extreme cold had actually kept her alive.
>
> "Over the last 28 years, there have been 34 victims of accidental hypothermia with cardiac arrest who were rewarmed on cardiopulmonary bypass and 30% survived," he said. "The key question is: are you cooled before you have the cardiac arrest or are you first having a circulatory arrest and then getting cooled?"

Cox goes on to explain some of the key biology involved:

While lowering the body temperature will stop the heart, it also reduces the oxygen demand of the body and, in particular, the brain cells. If the vital organs have been sufficiently cooled before the cardiac arrest occurs, then the inevitable cell death from the lack of circulation will be postponed, buying emergency services an extra time window to try and save the person's life.

"Hypothermia is so fascinating because it's a double-edged sword," Gilbert said. "On the one side it can protect you but, on the other side, it will kill you. But it's all a question of how controlled the hypothermia is. Anna was probably cooled quite slowly but efficiently so that, when her heart stopped, her brain was already so cold that the oxygen need in the brain cells was down to zero. Good CPR can provide up to 30-40% of the blood circulation to the brain and in these cases that is often sufficient to keep the person alive for sometimes seven hours while we try to restart the heart."

Thankfully, Bågenholm made almost a complete recovery. Ten years later, she was working as a radiologist in the hospital where her life had been saved. Bågenholm had experienced accidental hypothermia. Increasingly, doctors are deliberately inducing hypothermia, to obtain time to carry out complicated medical procedures. In his book *Extreme Medicine*,[11] Kevin Fong tells the story of the treatment in 2010 of Esmail Dezhbod:[12]

Esmail Dezhbod's symptoms had begun to worry him. He felt pressure in his chest, at times great pain. A body scan revealed that Esmail was in trouble. He had an aneurysm of his thoracic aorta, a swelling of the main arterial tributary leading from his heart. This vessel had doubled in size, to the width of a can of Coke.

Esmail had a bomb in his chest that might go off at any moment. Aneurysms elsewhere can usually be repaired with relative ease. But in this location, so close to the heart, there are no easy options. The thoracic aorta carries blood from the heart and into the upper body, supplying oxygen to the brain, among other organs. To repair the aneurysm, flow would have to be interrupted by stopping the heart. At normal body temperatures, this and the accompanying oxygen starvation would damage the brain, leading to permanent disability or death within three or four minutes.

Esmail's surgeon, cardiac specialist John Elefteriades, MD, decided to carry out the procedure under the conditions of deep hypothermic arrest. He used a heart-lung bypass machine to cool Esmail's body to a mere 64.4 °F before stopping his heart completely. Then, while the heart and circulation were at a standstill, Dr. Elefteriades performed the complicated repair, racing the clock while his patient lay dying on the operating table…

It's a delicate operation:

Though Dr. Elefteriades is an old hand with hypothermic arrest, he says that every time feels like a leap of faith. Once circulation has come to a standstill, he has no more than about 45 minutes before irreversible damage to the patient's brain occurs. Without the induced hypothermia, he would have just four.

The doctor lays the stitches down elegantly and efficiently, making every movement count. He has to cut out the diseased section of the aorta, a length of around six inches, then replace it with an artificial graft. The electrical activity in Esmail's brain is, at this point, undetectable. He is not breathing and has no pulse. Physically and biochemically, he is indistinguishable from someone who is dead.

That phrase is worth emphasising: "Physically and biochemically, he is indistinguishable from someone who is dead". However, he is still capable of revival. Fong continues:

After 32 minutes, the repair is complete. The team warms Esmail's freezing body, and very quickly his heart explodes back to life, pumping beautifully, delivering a fresh supply of oxygen to his brain for the first time in over half an hour.

Fong reports that he visited the patient in the intensive care unit the following day: "He is awake and well. His wife stands by his bed, overjoyed to have him back."

Who would deny the patient's wife the chance to have that joyous reuniting with her husband? Nevertheless, the critics of cryonics would deny many other people the chance to anticipate a similarly joyous reuniting with their loved friends and families at the culmination of a cryonics suspension. They would say that the extrapolation from therapeutic hypothermia to cryonics is too big a chasm. The temperature involved in cryonics is much lower – the temperature of liquid nitrogen – and the timescale of the suspension is much longer. In response, we maintain there are promising grounds to believe this chasm can, indeed, be bridged.

Not Freezing

A second precursor that points the way to successful cryonics technology is the fact that some organisms can already survive various kinds of sub-zero hibernation. For example, the Artic ground squirrel hibernates for up to eight months each year, during which time its core temperature drops from 36 °C to -3 °C, whilst external temperatures can reach as low as -30 °C. *The New Scientist* reports that:[13]

> To prevent their blood from freezing, the squirrels cleanse it of any particles that water molecules could form ice crystals around. This allows the blood to remain liquid below zero, a phenomenon known as supercooling.

Various fish in polar regions can survive in salt water that is below the freezing point of fresh water. They seem to manage this, without their blood freezing, with the help of so-called antifreeze proteins (AFPs). AFPs suppress the growth of ice crystals. Species of insects, bacteria, and plants also take advantage of AFPs.[14] Remarkably, larvae of the Alaskan beetle have been reported as surviving temperatures as low as -150 °C, by means of adopting a glass-like vitrified state.[15]

The champion species for surviving ultra-low temperatures is the tardigrade, which is sometimes known as the "water bear". It's actually tiny: it grows to less than 2 mm. The species is also evolutionarily ancient, having been in existence some 500 million years ago, in the Cambrian era. An article on *BBC Earth* describes their tolerance of temperatures actually below that of liquid nitrogen as used in cryonics (-196 °C). It refers to experiments carried out in the 1920s by Gilbert Franz Rahm, a Benedictine monk:[16]

> Rahm… immersed [tardigrades] in liquid air at -200 °C for 21 months, in liquid nitrogen at -253 °C for 26 hours, and in liquid helium at -272 °C for 8 hours. Afterwards the tardigrades sprang back to life as soon as they came into contact with water.
>
> We now know that some tardigrades can tolerate being frozen to -272.8 °C, just above absolute zero… The tardigrades coped with a profound chill that does not occur naturally and must be created in the lab, at which atoms come to a virtual standstill.
>
> The biggest hazard tardigrades face in the cold is ice. If ice crystals form inside their cells, they can tear apart crucial molecules like DNA.
>
> Some animals, including some fish, make antifreeze proteins that lower the freezing point of their cells, ensuring that ice doesn't form. But these proteins haven't been found in tardigrades.
>
> Instead it seems tardigrades can actually tolerate ice forming within their cells. Either they can protect themselves from the damage caused by ice crystals, or they can repair it.
>
> Tardigrades may produce chemicals called ice nucleating agents. These encourage ice crystals to form outside their cells rather than inside, protecting

the vital molecules. Trehalose sugar may also protect those that produce it, as it prevents the formation of large ice crystals that would perforate the cell membranes.

The *C. elegans* nematode worm, whose variable longevity has featured in many of the experiments covered in previous chapters in this book, makes an important appearance in this chapter too. On this occasion, what's noteworthy is the preservation of memories over the process of *C. elegans* individuals being cryonically suspended (to the temperature of liquid nitrogen) and then reanimated. The experiment was carried out by Natasha Vita-More of the University of Advancing Technology, Tempe, Arizona, and Daniel Barranco of Universidad de Sevilla, Spain. Here's the description of the experiment from the abstract of their October 2015 article entitled "Persistence of Long-Term Memory in Vitrified and Revived *Caenorhabditis elegans*" and published in *Rejuvenation Research:*[17]

Can memory be retained after cryopreservation? Our research has attempted to answer this long-standing question by using the nematode worm Caenorhabditis elegans, a well-known model organism for biological research that has generated revolutionary findings but has not been tested for memory retention after cryo-preservation. Our study's goal was to test C. elegans' memory recall after vitrification and reviving. Using a method of sensory imprinting in the young C. elegans, we establish that learning acquired through olfactory cues shapes the animal's behaviour and the learning is retained at the adult stage after vitrification. Our research method included olfactory imprinting with the chemical benzaldehyde for phase-sense olfactory imprinting at the L1 stage, the fast-cooling SafeSpeed method for vitrification at the L2 stage, reviving, and a chemotaxis assay for testing memory retention of learning at the adult stage. Our results in testing memory retention after cryopreservation show that the mechanisms that regulate the odorant imprinting (a form of long-term memory) in C. elegans have not been modified by the process of vitrification or by slow freezing.

In an article co-authored by Vita-More in *MIT Technology Review*, "The Science Surrounding Cryonics", the significance of this *C. elegans* result is put in context. The question under discussion is whether there is any possibility that human memory and consciousness could survive cryonic suspension. Vita-More and colleagues write as follows:[18]

The exact molecular and electrochemical features of the brain that underlie the conscious mind remain far from completely explored. However, available evidence lends support to the possibility that brain features that encode memories and determine behaviour can be preserved during and after cryopreservation.

Cryopreservation is already used in laboratories all over the world to maintain animal cells, human embryos, and some organized tissues for periods as long as three decades. When a biological sample is cryopreserved, cryoprotective chemicals such as DMSO or propylene glycol are added and the temperature of the tissue is lowered to below the glass transition temperature (typically about −120 °C). At these temperatures, molecular activities are slowed by more than 13 orders of magnitude, effectively stopping biological time.

Although no one understands every detail of the physiology of any cell, cells of virtually every conceivable kind are successfully cryopreserved. Similarly, while the neurological basis for memory, behaviour, and other features of a person's identity may be staggeringly complex, understanding this complexity is a problem largely independent of being able to preserve it.

Vita-More and colleagues then highlight the evidence from *C. elegans* that memories can survive cryopreservation:

For decades *C. elegans* have commonly been cryopreserved at liquid nitrogen temperatures and later revived. This year, using an assay for memories of long-term odorant imprinting associations, one of us published findings that C. elegans retain learned behaviours acquired before cryopreservation. Similarly, it has been shown that long-term potentiation of neurons, a mechanism of memory, remains intact in rabbit brain tissue following cryopreservation.

Reversibly cryopreserving large human organs, such as hearts or kidneys, is more difficult than preserving cells but is an active area of research with important public health benefits, since it would greatly increase the supply of organs for transplant. Researchers have made progress in this area, successfully cryopreserving and later transplanting sheep ovaries and rat limbs, and routinely recovering rabbit kidneys after cooling to −45 °C. Efforts to improve these technologies provide indirect support for the idea that the brain, like any other organ, may be adequately cryopreserved by current methods or methods under development.

Note that the cryonicists are very clear that the preservation methods they use should be described as "vitrification" rather than "freezing". The difference is explained straightforwardly, with easily understood graphics, on the website of *Alcor*, one of the leading providers of cryonics services. Here's the key conclusion:[19]

Because no ice is formed, vitrification can solidify tissue without structural damage.

Given that point, it's (almost) remarkable that various high-profile critics of cryonics seek to discredit the whole concept by theatrically demonstrating the structural damage done to fruit and vegetables – such as strawberries and carrots – when they are frozen and then thawed. The critics almost sneer: *how could the cryonicists be so stupid?* We are tempted to sneer in return: *how could these critics get their basic facts so badly wrong?* Are these critics really unaware of the successful cryopreservation of human embryos (pivotal in IVF treatments)? Have they not heard about the 2002 vitrification by Greg Fahy and colleagues at twenty-first-Century Medicine of a rabbit kidney, which was lowered to -122 °C before being thawed and successfully transplanted, as an operational organ, into another rabbit?[20]

As we have seen, there's more going on here than rational debate. It's another example of a gulf of two paradigms, with pressures of adverse psychology making it hard for some observers to take seriously the possibility of cryonics. The possibility that cryonics might work poses a strong threat to the framework of ideas with which many people have surrounded themselves – the framework that says that "good people accept the inevitability of aging and death, and shouldn't fight that conclusion". People who have grown comfortable with that conclusion are therefore motivated to find fault with the cryonics worldview. That can explain why they blithely parrot technical objections, economic objections, or sociological objections which, frankly, don't hold up to serious scrutiny. As the British philosopher Max More explains:[21]

We'll look back on this 50 to 100 years from now — we'll shake our heads and say, "What were people thinking? They took these people who were very nearly viable, just barely dysfunctional, and put them in an oven or buried them under the ground, when there were people who could have put them into cryopreservation."

The Forthcoming Surge in Cryonics and Other Technologies

There's a great deal more than can be said about cryonics, from multiple viewpoints. It takes considerable time to sort through all the objections and misunderstandings that have grown up around cryonics. For an engaging introduction to the subject, we recommend the comprehensive March 2016 *Wait But Why* article "Why Cryonics Makes Sense" by Tim Urban.[22]

In turn, that article includes pointers to lots of additional material. Readers may also value the wealth of perspectives contained in the volume "Preserving

Minds, Saving Lives" that is available from the *Alcor* website.[23] For a detailed technical review, we recommend the comprehensive 2022 book by nanotechnology expert Robert Freitas, *Cryostasis Revival: The Recovery of Cryonics Patients through Nanomedicine*.[24]

Another sign of increased organizational maturity of the global cryonics community is the emergence of Tomorrow Biostasis, founded in Berlin in 2019 by German medical doctor Emil Kendziorra, who also had experience as an entrepreneur and as a cancer researcher before switching his attention to cryopreservation.[25] The group have one medical standby team in Berlin, another one (via a partner) in Amsterdam, and will soon have a third team in Zurich. Their companion organization the European Biostasis Foundation has opened a patient storage facility in Rafz, Switzerland.[26] The authors have attended Biostasis conferences in Switzerland most years from 2016 onward and can report impressive signs of progress.[27]

We now propose to make a few final remarks about cryonics and why we think it is not only feasible but also going to go a long way in the next few years:

- The economic costs for cryonic preservation, long-term storage, and (assuming all goes well) an eventual reanimation, can presently be met from a life insurance policy
- The economic costs for an individual cryonic patient could decline by orders of magnitude if the number of patients grows significantly; this is the familiar principle of benefiting from "the economics of scale"
- As long as the "accepting aging" paradigm remains so pervasive within society, most people will feel strong social and psychological pressure against investigating cryonics and, subsequently, signing the relevant contracts. However, as this paradigm wilts (as we believe will happen) under greater publicity being given to rejuveneering breakthroughs, greater numbers of people will become open to the possibility of cryonics
- The increased interest in the subject will also result in more people carrying out research into improvements in cryonics – including improvements in the technology, engineering, support networks, business models, organisational frameworks, and methods to communicate about the subject to wider audiences. In turn, the innovations that result will accelerate the attractiveness of the cryonics option
- As high-profile figures from fields such as entertainment, business, academia, and the arts increasingly endorse the idea, it will open the way for wider members of the general public to feel comfortable about identifying themselves as cryonicists.

But cryopreservation is by no means the only idea whereby people might be transported (so to speak) from the present BR (Before Rejuvenation) era into AR (After Rejuvenation). Cryonics will probably continue to spread throughout the world, especially now that we are so close to reversing aging. We are facing with the last generation of mortals and the first generation of immortals. People will not want to die and be cremated or buried when they know that there are alternatives, albeit still with low probability, to reanimate them in the future.

The main "radical alternative" discussed here is cryopreservation, but it is not the only possibility that the future will offer us. What motivates cryopreservation is the possibility that, at some point, medicine will have advanced so far that extremely potent rejuvenation therapies are available. The use of these future therapies would cure patients of whatever was about to end their lives before they were suspended through cryonics. In principle, the use of these therapies would return patients to an excellent state of health. In the meantime, we believe it will be relatively affordable to keep them in liquid nitrogen indefinitely. Furthermore, besides cryopreservation, a few scientists are also researching other possibilities like chemopreservation and different types of plastination. These methods have their own issues and challenges, but have their strong supporters.[28]

Scientists are also conducting experiments to conserve the brain in other ways. We believe that the fundamental thing is to preserve the structure of the synapses at the moment the person dies. It is even possible that we could read the contents of brain connections by other methods and technologies, before the person dies. There are already devices that capture information from more than 500 individual neurons, a number that will continue to grow at an exponential rate.

From a computational point of view, we are just beginning to understand the complexity of the human brain. Our brain, which contains nearly one hundred billion neurons, is the most complex structure in the universe known to date. However, scientists are working on the creation of artificial brains. They estimate that in two or three decades, we will be able to create more complex structures than the human brain. Thanks to Kurzweil's Law of Accelerating Returns (a more comprehensive version of Moore's Law), which describes the exponential growth of computer power, it is forecast that artificial intelligence will pass the Turing Test in 2029 and reach "technological singularity" in 2045. At that point, it will be impossible to distinguish between artificial intelligence and human intelligence. Then it will also be possible to upload all knowledge, memories, experiences, and feelings to computers or the Internet (the "cloud"), which will even have an expansible and superior memory to human one.

Artificial memory will also improve and grow, as will the capacity and processing speed of artificial intelligence. It will all be part of an accelerated process of improving human intelligence thanks to continuous technological evolution. Humanity is just beginning to tread the fascinating path from biological evolution to technological evolution, a new conscious and intelligent evolution. According to Kurzweil, one kilogram of *computronium* (the hypothetical maximum unit of computation of matter) has the theoretical capacity to process about 5×10^{50} operations per second, something that acquires its true dimension if we compare it with a human brain that can process between 10^{17} and 10^{19} operations per second (according to different estimates). So, we still have enormous potential to develop, over several orders of magnitude, to further increase human and later post-human intelligence, moving from conventional biological brains to augmented post-biological brains. All this is part of the ideas of extension and expansion of life. Thus, concludes Kurzweil in his book *How to Create a Mind*.[29]

Waking up the universe, and then intelligently deciding its fate by infusing it with our human intelligence in its nonbiological form, is our destiny.

Notes

1. https://cryonics.org/cryonics-library/the-prospect-of-immortality/
2. http://www.bbc.com/future/story/20140821-i-will-be-frozen-when-i-die
3. https://cryonics.org/patient-details/
4. https://www.alcor.org/about/
5. https://www.longecity.org/forum/page/index.html/_/articles/cryonics
6. https://www.kurzweilai.net/playboy-reinvent-yourself-the-playboy-interview
7. https://www.biostasis.com/scientists-open-letter-on-cryonics/
8. http://cryonics-research.org.uk/
9. http://www.thelancet.com/journals/lancet/article/PIIS0140-6736(0 0)01021-7/
10. https://www.theguardian.com/science/blog/2013/dec/10/life-death-therapeutic-hypothermia-anna-bagenholm
11. https://www.amazon.com/Extreme-Medicine-Exploration-Transformed-Twentieth/dp/1594204705
12. https://web.archive.org/web/20210411044332/https://www.rd.com/article/hypothermia-cheat-death/

13. https://www.newscientist.com/article/dn23107-zoologger-supercool-squirrels-go-into-the-deep-freeze/
14. https://www.sciencedaily.com/releases/2011/04/110411152533.htm
15. http://jeb.biologists.org/content/213/3/502.full
16. http://www.bbc.co.uk/earth/story/20150313-the-toughest-animals-on-earth
17. http://www.ncbi.nlm.nih.gov/pmc/articles/PMC4620520/
18. https://www.technologyreview.com/s/542601/the-science-surrounding-cryonics/
19. http://www.alcor.org/Library/html/vitrification.html
20. http://www.bbc.com/future/story/20140224-can-we-ever-freeze-our-organs
21. http://www.kurzweilai.net/alcor-update-from-max-more-new-ceo
22. http://waitbutwhy.com/2016/03/cryonics.html
23. http://www.alcor.org/book/index.html
24. https://www.alcor.org/cryostasis-revival/
25. https://www.tomorrow.bio/
26. https://ebf.foundation/
27. https://www.biostasis2022.com/
28. https://www.brainpreservation.org/faq-items/17-what-problems-currently-exist-for-chemopreservation/
29. https://www.amazon.co.uk/How-Create-Mind-Ray-Kurzweil/dp/0715647334

9

The Future Depends on Us

The rapid progress true science now makes, occasions my regretting sometimes that I was born so soon. It is impossible to imagine the height to which may be carried in a thousand years the power of man over matter [...] All diseases may, by sure means, be prevented or cured, not even excepting that of old age, and our lives lengthened at pleasure even beyond the antediluvian standard.

Benjamin Franklin, 1780

Now this is not the end. It is not even the beginning of the end. But it is, perhaps, the end of the beginning.

Winston Churchill, 1942

We want to live forever, and we're getting there.

Bill Clinton, 1999

Yes, I was singled out for death; no, I'm not actually planning to die.

Sergey Brin, 2017

The rejuveneering project has made a great deal of progress over the last three decades. Aging is understood much better today than in any previous era. What's more, as previous chapters have shown, there are many grounds for anticipating an acceleration of progress over the next two or three decades. This progress should see the creation of practical bioengineering therapies that will take increasing advantage of our expanding theoretical knowledge. There are credible scenarios ahead, for reaching a state of affairs by around 2040 in which the terrible diseases of aging have become as rare as, say, polio and smallpox are today.

Nevertheless, many uncertainties lie ahead. These aren't just uncertainties in detail—over, for example, which drug will prove to have the biggest short-term impact on healthy lifespan, or which AI algorithm will deliver the most important insights into modifications in gene pathways. Instead, these are uncertainties over fundamentals—problems that could jeopardise the entire rejuveneering project.

It's time to look more closely at what could be the biggest obstacles on the pathway to the abolition of aging. Out of all the questions that audiences ask us when we talk to them about the potential for rejuveneering, these are the hardest to answer.

Exceptional Engineering Complications?

Sometimes, problems prove much harder to solve than people expected. Consider nuclear fusion. It's commonly said that nuclear fusion is always thirty years in the future. An article by Nathaniel Scharping in *Discover*, "Why Nuclear Fusion Is Always 30 Years Away", summarises the experiences of the nuclear fusion industry:[1]

> Nuclear fusion has long been considered the "holy grail" of energy research. It represents a nearly limitless source of energy that is clean, safe and self-sustaining. Ever since its existence was first theorized in the 1920s by English physicist Arthur Eddington, nuclear fusion has captured the imaginations of scientists and science-fiction writers alike. Fusion, at its core, is a simple concept. Take two hydrogen isotopes and smash them together with overwhelming force. The two atoms overcome their natural repulsion and fuse, yielding a reaction that produces an enormous amount of energy.
>
> But a big payoff requires an equally large investment, and for decades we have wrestled with the problem of energizing and holding on to the hydrogen fuel as it reaches temperatures in excess of 150 million degrees Fahrenheit…
>
> The most recent advancements have come from Germany, where the Wendelstein 7-X reactor recently came online with a successful test run reaching almost 180 million degrees, and China, where the EAST reactor sustained a fusion plasma for 102 seconds, although at lower temperatures.
>
> Still, even with these steps forward, researchers have said for decades that we're still 30 years away from a working fusion reactor. Even as scientists take steps toward their holy grail, it becomes ever more clear that we don't even yet know what we don't know.

The problem is that each step forward seems to throw up new issues which are just as hard to solve as the previous ones, that is, for every answer, more questions:

The Wendelstein 7-X and EAST reactor experiments were dubbed "break-throughs," which is an adjective commonly applied to fusion experiments. Exciting as these examples may be, when considered within the scale of the problem, they are only baby steps. It is clear that it will take more than one, or a dozen, such "breakthroughs" to achieve fusion.

"I don't think we're at that place where we know what we need to do in order to get over the threshold," says Mark Herrmann, director of the National Ignition Facility in California. "We're still learning what the science is. We may have eliminated some perturbations, but if we eliminate those, is there another thing hiding behind them? And there almost certainly is, and we don't know how hard that will be to tackle."

Might a similar set of ever harder problems lie ahead of the rejuveneering project? Perhaps each new tweak to the human biology, that enhances some aspect of healthy longevity, will have its own drawbacks. For example, we might strengthen the immune system, but this enhancement might cause the immune system to attack cells which the body needs for its normal healthy functioning—similar to the way that type 1 diabetes can result from an over-aggressive immune system destroying the islet cells in the pancreas which would otherwise produce insulin. And a second engineering intervention to fend off that unintended side-effect might, in turn, generate yet further complications. Similarly, lengthened telomeres might cause an increase in the incidence of cancer. Although not likely, it is possible.

One reason to be doubtful that any such fundamental engineering impasse lies ahead is because we can already see other animals—including some that experience negligible senescence—that can have much longer lifespans than humans. Nevertheless, it's possible in principle that our unique human attributes might somehow get in the way of engineering modifications that would provide negligible senescence for us. In ways that we don't yet understand, it could be the case that rejuveneering will suffer the same fate as nuclear fusion, with its advent being repeatedly delayed.

After all, sometimes a problem that is comparatively easy to state can require enormous processing to solve. The mathematical problem known as Fermat's Last Theorem is one example. The theorem was stated in the margin of a copy of a textbook by Pierre de Fermat in 1637. It's very short: "the equation $a^n + b^n = c^n$ has no solutions in positive integers, if n is an integer greater than 2". Nevertheless, the theorem took the entire mathematical community a total of 358 years to prove. The proof, by Andrew Wiles, occupied over 120 pages when it was published in two articles in the *Annals of Mathematics* in 1995, including nearly 10 pages of references to previous mathematical papers.[2] This centuries-long saga of development would surely have shocked

Fermat, if he could have foreseen it; in fact Fermat had convinced himself that he already had worked out a proof of the theorem, that was, however, too long for him to write into that same margin.

Despite the possible comparisons to nuclear fusion and to Fermat's Last Theorem, we think it's unlikely that insoluble engineering hurdles lie ahead on the rejuveneering footpath. It's not as if there is only one engineering technique that can be investigated. On the contrary, numerous different types of rejuveneering intervention can be considered.

Moreover, the slow progress with nuclear fusion can be attributed to factors other than sheer technical difficulty. In his *Discovery* article, Scharping states that the fusion project has lacked sufficient funding, and is being held up by the political difficulties of international cooperation.

More than a scientific problem

Ultimately, the question may be one of funding. Multiple sources said they were confident that their research could progress faster if they received more support. Funding challenges certainly aren't new in scientific research, but nuclear fusion is particularly difficult due to its near-generational timescale. Although the potential benefits are apparent, and would indeed address issues of energy scarcity and environmental change that are relevant today, the day when we see a payoff from fusion research is still far in the future.

Our desire for an immediate return on our investments dampens our enthusiasm for fusion research, says Laban Coblentz, the head of Communication at ITER.

"We want our football coaches to perform in two years or they're out, our politicians have two or four or six years and they're out — there's very little time to return on investment," he said. "So when somebody says we'll have this ready for you in 10 years, that's a tough narrative to tell."

In the U.S., fusion research receives less than $600 million in funding a year, including our contributions to ITER. This is a relatively small sum when compared to the $3 billion the Department of Energy requested for energy research in 2013. Overall, energy research represented 8% of the total funding the U.S. gave out for research that year.

"If you look at it in terms of energy budgets, or what's spent on military development, it's not really a lot of money that's going to this," says Thomas Pedersen, division head at the Max-Planck Institut für Plasmaphysik. "If you compare us to other research projects, it seems very expensive, but if you compare it to what goes into oil production or windmills or subsidies for renewables, its much, much less than that."

Sharping concludes that the progress of nuclear fusion will come down to a question of political will:

Fusion power is always 30 years away.

However, the finish line has been visible for some time now, a mountaintop that seems to recede with every step forward. It is the path that is obscured, blocked by obstacles that are not only technological, but also political and economic in nature. Coblentz, [Hutch] Neilson and [Duarte] Borba expressed no doubts that fusion is an achievable goal. When we reach it however, may be largely dependent on how much we want it.

Soviet physicist, Lev Artsimovich, the "Father of the Tokamak" may have summed it up best:

"Fusion will be ready when society needs it."

In this aspect, the comparison between fusion and rejuveneering is actually an apt one:

- The engineering challenges are deeply hard, in both cases, but are by no means insoluble
- Progress in solving these challenges will depend upon large international collaboration, backed by political support of the sort we'll cover later in this chapter
- The speed at which this large international collaboration can be created and supported depends, in turn, on the level of public demand for a solution
- Both endeavours are likely to be accelerated to a swifter conclusion by leaps forward in the power of artificial intelligence.[3]

As an aside, we suspect that, if the survival of the human race had been manifestly dependent upon someone finding a proof of Fermat's Last Theorem, such a proof could have been found a lot more quickly than actually happened. A wartime siege mentality mindset can work wonders—*so long as there still exists an infrastructure adequate to support the collaboration of brilliant minds.*

Market Failures?

The need for smart regulations and, more generally, for informed state guidance over technological development, is underscored by a number of observations. What these observations have in common is that an economic free market, if left to itself, can sometimes produce outcomes that are far from optimal—and, indeed, can be disastrous.

One example is the way that pharmaceutical companies routinely deprioritise the development of drugs for diseases that only impact populations that have low incomes. The organisation "Drugs for Neglected Diseases initiative" (DNDi) was set up in 2003 to address that issue. The DNDi website gives sobering details of some of these "neglected diseases":[4]

- Malaria—kills one child every minute in sub-Saharan Africa (about 1300 children every day)
- Paediatric HIV—2.6 million children below 15 years of age are living with HIV globally, mainly in sub-Saharan Africa, and 410 of them die every day
- Filarial Diseases—120 million people are infected with Elephantiasis and 25 million with River Blindness
- Sleeping sickness—endemic in 36 African countries with 21 million people at risk
- Leishmaniasis—occurs in 98 countries with 350 million people at risk worldwide
- Chagas disease—endemic in 21 countries across Latin America; kills more people in the region than malaria.

In summary:

Neglected diseases continue to cause significant morbidity and mortality in the developing world. Yet, of the 850 new therapeutic products approved between 2000 and 2011, only 4% (and only 1% of all approved NCEs [New Chemical Entities]) were indicated for neglected diseases, even though these diseases account for 11% of the global disease burden.

This situation shouldn't come as a surprise, given the shareholder constraints under which pharmaceutical companies operate. For example, the stated policy of pharmaceutical giant Bayer was described in an article by Glyn Moody in early 2014. The article carried the headline "Bayer's CEO: We Develop Drugs for Rich Westerners, Not Poor Indians". It quoted Bayer Chief Executive Officer Marijn Dekkers stating this principle:[5]

We did not develop this medicine for Indians. We developed it for western patients who can afford it.

That policy aligns with the for-profit motivation that the company pursues, in service of the needs of its shareholders to maximise returns. It's for that reason that DNDi advocate an "alternative model", stating their organizational vision as follows:

> To improve the quality of life and the health of people suffering from neglected diseases by using an alternative model to develop drugs for these diseases and by ensuring equitable access to new and field-relevant health tools.
>
> In this not-for-profit model, driven by the public sector, a variety of players collaborate to raise awareness of the need to research and develop drugs for those neglected diseases that fall outside the scope of market-driven R&D. They also build public responsibility and leadership in addressing the needs of these patients.

Glyn Moody, after noting the stark comments by Bayer CEO Dekkers mentioned above, points out how pharmaceutical companies have, in the past, shown broader motivation. He refers to this quote from 1950 from George Merck (emphasis added):[6]

> "We try never to forget that medicine is for the people. It is not for the profits. The profits follow, and if we have remembered that, they have never failed to appear. The better we have remembered it, the larger they have been...
>
> We cannot step aside and say that we have achieved our goal by inventing a new drug or a new way by which to treat presently incurable diseases, a new way to help those who suffer from malnutrition, or the creation of ideal balanced diets on a worldwide scale. *We cannot rest till the way has been found, with our help, to bring our finest achievement to everyone.*

What determines whether the narrow financial incentives of the market govern behaviours of companies with the technology (possibly unique technology) that enables significant human enhancement? Other factors need to come into play—not just financial motivation.

Even within their own parameters—the promotion of optimal trade and the accumulation of wealth—free markets sometimes fail. The argument for smart oversight and regulation of markets is well made in the 2009 book "How markets fail: the logic of economic calamities" by the *New Yorker* journalist John Cassidy.[7]

The book contains a sweeping but compelling survey of a notion Cassidy dubs "Utopian economics", before providing layer after layer of critique of that notion. As such, the book provides a useful guide to the history of economic thinking, covering Adam Smith, Friedrich Hayek, Milton Friedman, John Maynard Keynes, Arthur Pigou, Hyman Minsky, among others.[8]

The key theme in that book is that markets do fail from time to time, potentially in disastrous ways, and that some element of government oversight and intervention is both critical and necessary, to avoid calamity. This theme is hardly new, but many people resist it, and Cassidy's book has the merit of marshalling the arguments comprehensively.

As Cassidy describes it, "utopian economics" is the widespread view that the self-interest of individuals and agencies, allowed to express itself via a free market economy, will inevitably produce results that are good for the whole economy. The book starts with eight chapters that sympathetically outline the history of thinking about utopian economics. Along the way, he regularly points out instances when free market champions nevertheless described cases when government intervention and control was required. Next, Cassidy devotes another eight chapters to reviewing the history of criticisms of utopian economics. This part of the book is entitled "Reality-based economics", and covers topics such as:

- Game theory ("the prisoners dilemma")
- Behavioural economics (pioneered by Daniel Kahneman and Amos Tversky)—including disaster myopia
- Problems of spillovers and externalities (such as pollution)—which can only be fully addressed by centralised collective action
- Drawbacks of hidden information and the failure of "price signalling"
- Loss of competiveness when monopoly conditions are approached
- Flaws in banking risk management policies (which drastically underestimated the consequences of larger deviations from "business as usual")
- Problems with asymmetric bonus structure
- The perverse psychology of investment bubbles.

These factors all obstruct markets from discovering the optimal solutions. In summary, Cassidy lists four "illusions" of utopian economics:

1. The illusion of harmony: that free markets always generate good outcomes
2. The illusion of stability: that free market economy is sturdy
3. The illusion of predictability: that distribution of returns can be foreseen
4. The illusion of Homo Economicus: that individuals are rational and act on perfect information.

These illusions remain pervasive in many parts of economic thought. These illusions also lie behind technolibertarian optimism that technology, without government intervention, will be able to solve social and climatic problems such as terrorism, surveillance, environmental devastation, extreme fluctuations in weather, threats from new pathogens, and the growing costs of diseases of old age.

Indeed, free markets and innovative technology have, together, been a tremendous force for progress in recent history. However, they need smart oversight and regulation if they are going to reach their fullest potential. Indeed, without such oversight and regulation, they may lead society into a new Dark Age, rather than an age of sustainable abundance and healthy longevity for all.

Poor Ways of Doing Good?

For readers for whom the discussion of politics in the last few sections extended beyond their comfort zone, in this section the narrative transitions away from politics, into an area that might be best described as "philosophy".

One of the biggest threats to the rejuveneering project is that muddled thinking will prevail, in the public mind, about what kind of actions are admirable. People who wish to behave in an admirable way may, nevertheless, be guided by ideas that end up doing harm rather than good. As victims of social and psychological pressure, they'll be stuck, consciously or unconsciously, in the accepting aging paradigm. Their personal philosophies will lead them to take actions that actually cause damage, to themselves and to their fellow citizens.

Specifically, if people are convinced that it's praiseworthy to accept ongoing aging and impending death, as some kind of "natural order of things", they'll be inclined to oppose measures that would enable radically extended healthspans. Consciously or unconsciously, they'll (wrongly) see such measures as somehow unfair, or unbalanced, or disproportionate, or grasping, or egotistical, or juvenile.

People locked into that mindset will prefer that society invests its discretionary time and effort into projects that accept aging as a given. For example, they may support projects that seek to help the elderly by providing them with neighbourly contact, lower cost transport, or improved "assisted living" facilities. Other projects they may be comfortable to support would be that more people can live long enough to become elderly, instead of being stricken by accidents or diseases in their youth or middle age. Or they'll support an expansion of education for people at all ages. They'll see all these projects as admirable, acceptable ways of doing good. But they'll be blinded to the possibility that there could be a *better way for them to do good*.

The phrase "*Doing good better*" forms the title of a book written in 2015 by William MacAskill, who at the age of 28 was one of the youngest professors at Oxford University. The book's subtitle is "*Effective Altruism and how you can make a difference*". On his website, MacAskill introduces the book as follows:[9]

> Do you care about making the world a better place? Perhaps you buy ethical products, donate to charity or volunteer your time in the name of doing good. But how often do you know what impact you really have?
>
> In my book, I argue that many ways of making a difference achieve little, but that, by targeting our efforts on the most effective causes, we each have an enormous power to make the world a better place.

Some people find this kind of cold calculation to be unsettling. It can seem somewhat dehumanised. But the Effective Altruism advocates make a strong case that, by failing to think through these kinds of considerations, we would be falling short of our own potential for improving the human condition. If our aim is truly to improve the human condition—rather than just to *feel good* about *gestures* we make with the implied aim at improving the human condition—then we need to be able to rethink our priorities.

Any such rethink should weigh up the possibility that extending healthy lifespans, by the abolition of aging, might turn out to be an even more cost-effective intervention, when we recognise that the increase in DALYs (Disability Adjusted Life Year) from successful rejuvenation therapies would be very considerable indeed.

Aubrey de Grey made a similar argument in a presentation at Oxford in 2012, "The cost-effectiveness of anti-aging research":[10]

- If we truly care about preventing deaths, we should pay close attention to the factor which is responsible for around two thirds of all deaths worldwide, namely aging (note that the figure includes all deaths from aging-related diseases—deaths that would not occur in the absence of aging)
- That high fraction (which rises to over 90% in the industrialised world) makes aging "unequivocally the world's most serious problem"
- The importance of abolishing aging increases even further, when we additionally consider the many years of declining functionality and increasing disability that precede deaths from aging
- Treatments that slow aging will have the benefit of delaying frailty and the onset of diseases of aging; treatments that, additionally, repair the bodily and cellular damage caused by aging have the potential to indefinitely prevent frailty and the diseases of aging—thereby further increasing the expected DALYs metric

- The costs required to make significant progress with rejuvenation therapies don't need to be particularly huge; a budget of around $50 million per year, over 5–10 years, could well be enough to advance the rejuveneering therapies proposed by SENS to the point where they can be applied to middle-aged mice with dramatic effect

- Once middle-aged mice, who had previously not received any special treatment, have their remaining healthy lifespan increased by upwards of 50% following the administration of rejuvenation therapies, lots of other funding would quickly follow: governments, businesses, and philanthropists would by that time all understand and acknowledge the great potential of these therapies for humans too.

The task that is urgent, de Grey argues, is to carry out intelligent advocacy for the needed research budgets in the shorter-term—up to the point when a manifest demonstration of robust mouse rejuvenation causes a wholesale change in public mindset. This shorter-term intelligent advocacy can build momentum, once more people take the time to think things through dispassionately, and perhaps employ the conceptual methods of Effective Altruism. But it's still going to take a lot of effort—and a lot of smart marketing   to overcome the public's deep-rooted "accepting aging" apathy.

Public Apathy?

Broadly speaking, there are two approaches to overcome apathy and to change the world. Either you change the world directly, or you change people's minds about the importance of changing the world (so that one of them changes the world instead). In other words, either you get involved in actually doing things, or you talk about how good it would be if people did various things.

The first approach involves action. The second involves ideas. The first approach can be adopted by engineers, entrepreneurs, designers, and so forth. The second approach is, in principle, available to everyone—everyone who can speak up about the importance of an idea.

We are fans of both approaches, but we recognize the second has come under a lot of criticism. In an age of instant messaging, with legions of people who can click an online "Like" button whilst still in pyjamas or lounging on their sofas, it has become fashionable to decry so-called "*slacktivism*" (also known, less pithily, as "armchair activism"). Critic Evgeny Morozov, in his NPR article "Brave New World of Slacktivism", was withering in his scorn of the practice:[11]

"Slacktivism" is an apt term to describe feel-good online activism that has zero political or social impact. It gives those who participate in "slacktivist" campaigns an illusion of having a meaningful impact on the world without demanding anything more than joining a Facebook group. Remember that online petition that you signed and forwarded to your entire contacts list? That was probably an act of slacktivism.

"Slacktivism" is the ideal type of activism for a lazy generation: why bother with sit-ins and the risk of arrest, police brutality, or torture if one can be as loud campaigning in the virtual space? Given the media's fixation on all things digital—from blogging to social networking to Twitter—every click of your mouse is almost guaranteed to receive immediate media attention, as long as it's geared towards the noble causes. That media attention doesn't always translate into campaign effectiveness is only of secondary importance…

The real issue here is whether the mere availability of the "slacktivist" option is likely to push those who in the past might have confronted the regime in person with demonstrations, leaflets, and labour organizing to embrace the Facebook option and join a gazillion online issue groups instead. If this is the case, then the much-touted tools of digital liberation are only driving us further away from the goal of democratization and building global civil society.

In contrast to this negative assessment, we see a very important role for online advocacy, in the battle to raise public awareness of the profound opportunities of rejuveneering and the profound risks from the potential misuse of the same underlying technologies. For example, social networks had a major impact on government change in several countries across the Middle East and other parts of the world over the past decade.

Beyond social networks, more traditional media such as print, radio, and television remain important. Other forms of communication like film, music, books, lectures, poetry, and art are equally fundamental. Even YouTube videos can help mobilize people, as is the case with the superb "Why Age? Should We End Aging Forever?" which during its first four months saw more than four million people.[12] That's a far cry from the 4.6 billion times the video clip of the song "Despacito" has been seen on YouTube in its first year, but it's a lot better than nothing.[13]

It is also essential to turn the ideas of anti-aging and rejuvenation into viral ideas, including both the extension and expansion of human life. Ideally, memes should be created to help "viralize" the ideas of the new paradigm of indefinite disease-free youth, for the benefit would be incalculable for everyone, without discrimination.

Another way of communicating that is worth mentioning to reduce apathy is, for example, a fantastic short story by Swedish philosopher Nick Bostrom, a professor at Oxford University. We have already mentioned and praised *The Fable of the Dragon-Tyrant*, written in 2005. This fable analogizes the accepting aging paradigm to a centuries-long acquiescence by the citizens of a fictional country to the demands of a giant dragon:[14]

> It demanded from humankind a blood-curdling tribute: to satisfy its enormous appetite, ten thousand men and women had to be delivered every evening at the onset of dark to the foot of the mountain where the dragon-tyrant lived. Sometimes the dragon would devour these unfortunate souls upon arrival; sometimes again it would lock them up in the mountain where they would wither away for months or years before eventually being consumed…

That fable was converted into an engaging video format by CGP Grey in 2018, and it has already been viewed more than nine million times.[15]

Max More is another philosopher with an imaginative touch. We remember being struck by the thoughtfulness of his 1999 *Letter to Mother Nature*, which starts as follows:[16]

> Dear Mother Nature:
>
> Sorry to disturb you, but we humans—your offspring—come to you with some things to say. (Perhaps you could pass this on to Father, since we never seem to see him around.) We want to thank you for the many wonderful qualities you have bestowed on us with your slow but massive, distributed intelligence. You have raised us from simple self-replicating chemicals to trillion-celled mammals. You have given us free rein of the planet. You have given us a life span longer than that of almost any other animal. You have endowed us with a complex brain giving us the capacity for language, reason, foresight, curiosity, and creativity. You have given us the capacity for self-understanding as well as empathy for others.
>
> Mother Nature, truly we are grateful for what you have made us. No doubt you did the best you could. However, with all due respect, we must say that you have in many ways done a poor job with the human constitution. You have made us vulnerable to disease and damage. You compel us to age and die—just as we're beginning to attain wisdom. You were miserly in the extent to which you gave us awareness of our somatic, cognitive, and emotional processes. You held out on us by giving the sharpest senses to other animals. You made us functional only under narrow environmental conditions. You gave us limited memory, poor impulse control, and tribalistic, xenophobic urges. And, you forgot to give us the operating manual for ourselves!
>
> What you have made us is glorious, yet deeply flawed. You seem to have lost interest in our further evolution some 100,000 years ago. Or perhaps you have been biding your time, waiting for us to take the next step ourselves. Either way, we have reached our childhood's end.

We have decided that it is time to amend the human constitution.

We do not do this lightly, carelessly, or disrespectfully, but cautiously, intelligently, and in pursuit of excellence. We intend to make you proud of us. Over the coming decades we will pursue a series of changes to our own constitution, initiated with the tools of biotechnology guided by critical and creative thinking. In particular, we declare the following seven amendments to the human constitution:

Amendment No.1: We will no longer tolerate the tyranny of aging and death. Through genetic alterations, cellular manipulations, synthetic organs, and any necessary means, we will endow ourselves with enduring vitality and remove our expiration date. We will each decide for ourselves how long we shall live…

We reserve the right to make further amendments collectively and individually. Rather than seeking a state of final perfection, we will continue to pursue new forms of excellence according to our own values, and as technology allows.

Your ambitious human offspring.

Exponential Growth?

If done well, all the following can contribute to a seismic change in public mindset from being stuck in "accepting aging" to being receptive, and then fully supportive, of "anticipating rejuvenation": short videos, powerful online blogposts, soulful poems, eye-catching animations, witty limericks, clever jokes, dramatic performances, concept art, novellas, soaring anthems, chants, slogans, and evocative "memes" consisting of a picture and an associated memorable quote. All can help to dismantle public apathy and help to accelerate the growth of rejuvenation technologies worldwide.

And if slacktivists identify and highlight the best contributions from the many that are created, so that these contributions receive more attention, and hasten the weakening of the bastions of the accepting aging paradigm, that's something we heartily applaud. Once minds have been changed, actions can follow. When the groundwork is laid, new ideas can spread quickly.

What's hard, of course, is to know what time is the right time for a particular idea. If someone repeatedly cries wolf too early, they lose their credibility—and their audience. But we see plenty of reason why the present time is ripe for the idea that we can, and should, abolish aging. That idea can be backed up by a host of observations:

- Examples of animals that experience negligible senescence

- Genetic manipulations that can significantly extend lifespan (and healthspan)
- Fascinating possibilities from stem cell therapies
- The game-changing possibilities of CRISPR genetic editing
- The increasing viability of nano-interventions, such as nano-surgery and nano-bots
- Early indications that synthetic organs can be created
- Research projects targeting each of seven identified underlying causes of aging
- Encouraging progress in new ideas for treating cancer, as well as other diseases of aging
- Promising results from big data analysis by increasingly powerful artificial intelligence
- Financial models that show the tremendous economic benefits of the longevity dividend
- Examples from other technological fields of unexpectedly rapid progress
- Examples from other activist projects of rapid changes in social mindset.

These observations provide the environment in which the idea of the abolition of aging can thrive, but the task still remains to actually champion that idea:

- Finding better, more effective ways to express the idea, for different audiences
- Analysing the objections that people raise to the idea, and finding good responses to these objections
- Appreciating the underlying circumstances which make people want to object to the idea (or even just to ignore it), and taking steps, where possible, to transform these circumstances.

If these tasks get left undone, the idea may languish, being of interest only to a small minority. In that case, the accepting aging paradigm will remain dominant. Investment—both public and private—will go into fields other than rejuveneering. Regulatory hurdles will persist, that frustrate efforts by innovators to develop and deploy rejuvenation therapies. And upwards of 100,000 people will continue to die, every day, of diseases of aging that are actually avoidable. That would be the terrible cost of ongoing public apathy about the potential to abolish aging.

From the Abolition of Slavery in the Past to the Abolition of Aging in the Present

The abolition of slavery has a strong case to be one of the high points of human history. Drawing on the material in the magisterial book *"Inhuman Bondage: The Rise and Fall of Slavery in the New World"*[17] by veteran Yale historian David Brion Davis, Donald Yerxa of Boston University offers this assessment:[18]

> After receiving hundreds of antislavery petitions and debating the issue for years, the British Parliament passed the Abolition of the Slave Trade Act in March 1807. Starting May 1, 1807, no slaver could legally sail from a British port. Following the Napoleonic Wars, British abolitionist sentiment increased, and substantial public pressure was brought to bear on Parliament to gradually emancipate all British slaves. In August 1833, Parliament passed the Great Emancipation Act, which made provision for the gradual emancipation of slaves throughout the British Empire. Abolitionists on both sides of the Atlantic hailed it as one of the great humanitarian achievements in history. Indeed, the prominent Irish historian W.E.H. Lecky famously concluded in 1869 that "the unwearied, unostentatious, and inglorious crusade of England against slavery very may probably be regarded as among the three or four perfectly virtuous acts recorded in the history of nations."

> As the distinguished historian David Brion Davis observes, however, in his brilliant synthesis of slavery in the New World, British abolitionism is "controversial, complex, and even baffling." It has occasioned a significant historiographical debate lasting over sixty years. The key issue has been how to account for abolitionists' motives and the groundswell of public support for the antislavery cause. Davis suggests that historians find it difficult to accept that something as economically significant as the slave trade could be abolished on essentially religious and humanitarian grounds. After all, by 1805 "the colonial plantation economy," he informs us, "accounted for about one-fifth of Britain's total trade." Prominent abolitionists like William Wilberforce, Thomas Clarkson, and Thomas Fowell Buxton used Christian arguments to combat "inhuman bondage," but surely other, material factors were in play. A great deal of ink has been spilled assessing the relationship of antislavery to capitalism and free market ideology. And the upshot of this research is that the antislavery impulse went against British economic interests, both real and perceived.

So how do we explain the successes of a humanitarian movement advocating reforms that could have precipitated economic disaster? Davis concludes that while it is important to appreciate the complex interplay of economic, political, and ideological factors, we must recognize the significance of a moral vision that "could transcend narrow self-interest and achieve genuine reform."

The analysis by Davis makes it clear that:

- The abolition of slavery was by no means inevitable or predetermined
- There were strong arguments against the abolition of slavery—arguments raised by clever, devout people in both the United States and the United Kingdom—arguments concerning economic well-being, among many other factors
- The arguments of the abolitionists were rooted in a conception of a better way of being a human—a way that avoided the harsh bondage and subjugation of the slave trade, and which would in due course enable many millions of people to fulfil a much greater potential
- The cause of the abolition of slavery was significantly advanced by public activism—including pamphlets, lectures, petitions, and municipal meetings.

With its roots in the eighteenth century, and growing in momentum as the nineteenth century proceeded, the abolition of slavery eventually became an idea whose time had come—thanks to brave, smart, persistent activism by men and women with profound conviction. The American Civil War had much to do with slavery, which was finally eliminated in all states by President Abraham Lincoln in 1865. Thus, little by little, slavery was disappearing all over the world, until it was finally abolished in some Arab emirates during the 1960s, that is, more than a century and a half after the United Kingdom banned it.

With a different set of roots in the late twentieth century, and growing in momentum as the twenty-first century proceeds, the abolition of aging can, likewise, become an idea whose time has come. It's an idea about an overwhelmingly better future for humanity—a future that will allow billions of people to fulfil a much greater potential. But as well as excellent engineering—the creation of reliable, accessible rejuvenation therapies—this project will also require brave, smart, persistent activism, to change the public landscape from one hostile (or apathetic) to rejuveneering into one that deeply supports it.

Noise Swamping the Signal?

In our applause for activism in favour of rejuveneering, we don't mean to endorse all the rhetoric that can be found, online or in books, in favour of this project. Far from it. Indeed, much that is said in support of rejuveneering is probably counterproductive:

- Rash, unwarranted claims about the effectiveness of individual tonics or therapies
- Distortions of the research findings from particular experiments, with an eye to bolstering the market perception of products under commercial development
- Tiresome repetitions of misleading over-simplifications of more sophisticated principles
- Hurtful allegations about the competence or motivation of critical researchers who are actually carefully following established scientific processes
- Claims that are well understood to be false but which nevertheless keep being repeated, out of naivety and/or carelessness, by people who are well-intentioned but misinformed
- People being urged to undergo treatments which are actually dangerous.

The dangers of this kind of misrepresentation include various sorts of backlash:

- To protect patients from being misled and harmed, legislators may impose stricter regulation, that clamps down on good innovation as well as on purveyors of perceived "snake oil"
- Capable academics may want to dissociate themselves from the entire field, in order to avoid reputational damage
- Researchers may waste lots of time duplicating work that has already been done, and whose results ought to have been known in advance (but that knowledge was drowned out in the noise of low-quality communications)
- The public may tire of hearing of forthcoming rejuveneering therapies, and decide that the field is suspect and hype-laden
- Potential funding may be removed from the field, being routed instead to completely different kinds of project.

For these reasons, the rejuveneering community needs to work hard on improving its own knowledge management. Enthusiastic new members should be welcomed, but then be quickly brought up to speed in terms of the actual best state of current knowledge. They should be able to access online knowledge about:

- The community's best thoughts as to credible **roadmaps** for progress that can be made in the rejuveneering project in the years ahead
- The strengths and weaknesses of various **theories of aging**
- The **treatments and therapies** that are being developed or which are proposed
- The **lifestyle changes** which have the best chance, upon adoption, to keep individuals alive and healthy until such time as "Bridge 2" therapies become available
- The **history** of the overall field (to avoid needless repetition of previous mistakes)
- The broader **political**, **social**, **psychological**, and **philosophical** dimensions of rejuveneering
- The **projects** which are actively looking for assistance, and which the community judges to be worthy of support
- The **memes** of various sorts, at any given time, which are the most effective at winning new supporters and in responding to criticism
- The **skills** which are in short supply within the community, and the best ways in which various skills can be deployed in support of the rejuveneering goals
- The areas where **genuine differences of opinion** exist, and the proposed methods for how the community may be able to resolve these differences
- The **risks** which the community is tracking, and the proposed **mitigations** for these risks.

Evidently, the present book aims to cover many of the topics listed above. However, rejuveneering is a fast-changing field. Some of what we've written on these pages will be out-of-date, or otherwise incomplete, by the time you read it. For pointers to information that is more up-to-date and more comprehensive, please see online resources such as Lifespan.io ("crowdsourcing the cure for aging)",[19] Forever Healthy ("a private, humanitarian initiative with the mission of enabling people to vastly extend their healthy lifespan"),[20] and the German political party, Party for Biomedical Rejuvenation Research (slogan: "together against age-related diseases").[21]

To be clear, we are not saying that any new supporter of rejuveneering should be obliged to digest huge amounts of information before he or she is allowed to open their mouths in any public forum. The community's best knowledge about rejuveneering needs to be layered, easily searchable, and engaging. That way, when someone feels inspired to publicly address a particular topic, they should be able to quickly discover the community's best advice on what to say about it. They should also be able to find supportive, knowledgeable friendly people with whom they can discuss any issues arising. Whatever new insight arises from these conversations should be captured online, so that the knowledge base improves. As a result, the rejuveneering project can continue to move forwards.

Making a Real Difference?

In this chapter, we've reviewed some of the biggest risks facing the rejuveneering project. The project might become bogged down in enormous technical challenges that are harder than rejuveneers anticipate. It may alienate potential important supporters because of ill-chosen words and/or deeds, thereby cutting itself off from much-needed advice and finance.

Other latent support may fail to materialise, due to prevailing public apathy, with the accepting-aging paradigm remaining dominant. Yet other supporters could prove to be net hindrances to the project, magnifying confusion rather than actually helping.

Technoconservative politicians may put huge barriers in place of the research needed to create and deploy rejuvenation treatments. Technolibertarians might unwittingly precipitate an economic meltdown due to misguidedly dismantling public policies. Existential risks such as runaway climate change, highly virulent pathogens, or terrorist access to horrific weapons of mass destruction, could herald a terrible new dark age.

However, in this chapter, we've also indicated actions that can be taken by supporters of rejuveneering, to handle these risks, and to heighten the positive forces that sit alongside these negative risks. We ask each reader to consider which actions are the ones that best play to their own personal strengths.

Answers will vary from person to person. But we expect the following six types of action will feature prominently.

First, we need to strengthen our ties to **communities** that are working on at least parts of the rejuveneering project. We should find out which communities can nurture and inspire us, and where we can help, in turn, to nurture and inspire others. The resulting network ties will give us all greater strength to face the challenges ahead.

Second, we need to improve our personal **understanding** of aspects of rejuveneering—the science, roadmaps, history, philosophy, theories, personalities, platforms, open questions, and so on. With a better understanding, we can see more clearly what contributions we can make—and we can help others to make similar decisions for themselves. In some cases, we can help to document a better understanding of specific topics, by creating or editing knowledgebases or wikis.

Third, many of us can become involved with **marketing** of one sort or another. We might work on the creation and distribution of various marketing messages, presentations, videos, websites, articles, books, and so on. We might identify particular audiences—sets of people—and deepen the community's understanding of the issues high in the minds of these audiences. We might take the time to build better relationships with key influencers (potential new supporters of rejuveneering). We might even develop our political skillsets, improving our ability to influence others, forge alliances, broker coalitions, and create draft legislation in politician-friendly manner.

Fourth, some of us can undertake original **research**—into any of the unknowns of rejuveneering. This could be part of formal educational courses, or it could be a commercial R&D undertaking. It could also be part of a decentralised activity, in the style of "citizen science".[22]

Fifth, many of us can provide **funding** to projects that we judge to be particularly worthwhile. We can take part in specific fundraising initiatives, or we can donate some of our personal wealth. We can also decide to change our jobs, in order to earn more money, so we can make larger donations to the projects about which we care the most.

Last, but not least, we can work on our **personal effectiveness**—our ability to get things done. Having become aware of the historical importance of this present time period—a time period in which human society could make either a remarkable turn for the better or a remarkable turn for the worse—we should find ways to rise above the distractions and inertia of day-to-day "life as normal".

Instead of just being interested observers standing on the side lines of the culminating acts of humanity's oldest quest, occasionally offering cheers of encouragement, we can transform ourselves to become active participants in that quest. If we put our lives in order, each one of us can make a very real difference.

Notes

1. https://www.discovermagazine.com/technology/why-nuclear-fusion-is-always-30-years-away
2. http://www.jstor.org/stable/2118559

3. https://www.deepmind.com/blog/accelerating-fusion-science-through-learned-plasma-control
4. http://www.dndi.org/about-dndi/
5. https://www.techdirt.com/articles/20140124/09481025978/big-pharma-ceo-we-develop-drugs-rich-westerners-not-poor.shtml
6. https://todayinsci.com/M/Merck_George/MerckGeorge-Quotations.htm
7. https://www.newyorker.com/contributors/john-cassidy
8. http://www.amazon.com/How-Markets-Fail-Economic-Calamities/dp/0374173206/
9. https://www.williammacaskill.com/book
10. https://www.youtube.com/watch?v=jDJ_IjMwT20
11. http://www.npr.org/templates/story/story.php?storyId=104302141
12. https://www.youtube.com/watch?v=GoJsr4IwCm4
13. https://www.youtube.com/watch?v=kJQP7kiw5Fk
14. http://www.nickbostrom.com/fable/dragon.html
15. https://www.youtube.com/watch?v=cZYNADOHhVY
16. http://strategicphilosophy.blogspot.com/2009/05/its-about-ten-years-since-i-wrote.html
17. https://www.amazon.com/Inhuman-Bondage-Rise-Slavery-World/dp/0195339444
18. http://www.bu.edu/historic/london/conf.html
19. https://www.lifespan.io/
20. https://www.forever-healthy.org/
21. https://verjuengungsforschung.de/about_us
22. https://en.wikipedia.org/wiki/Citizen_science

Conclusion

Nothing is stronger than an idea whose time has come.
Victor Hugo, 1877

Whether you think you can, or you think you can't–you're right.
Henry Ford, 1946

Whether we abolish cancer, heart attacks, and dementia is no longer a question of if. It is a question of when.
Michael Greve, 2021

The Time Has Come

We are living fascinating times. Times of exponential change, times of total disruption, an incomparable period perhaps in the entire history of humanity. We are between the last mortal human generation and the first immortal human generation. The time has come to publicly declare the death of death. The alternative is very clear: if we do not kill death, death will kill us.

This is a call for the most important revolution in history, a revolution against aging and death, the great dream of all our ancestors. Aging has been and remains the greatest enemy of all humanity; it is the common enemy that we must defeat.

Unfortunately, until now, we had not had the science and technology to overcome aging. For the first time on the long and slow path of biological evolution, from our humble origins as small unicellular organisms billions of years ago, we can finally see the light at the end of the tunnel in this race for

© The Author(s), under exclusive license to Springer Nature Switzerland AG 2023
J. Cordeiro, D. Wood, *The Death of Death*, Copernicus Books,
https://doi.org/10.1007/978-3-031-28927-9

life. We are in a war against death, a war for life and our weapons are science and technology.

In 1861, in the midst of the continuous European wars of the nineteenth century, the French writer Gustave Aimard expressed the following thought in his novel *The Freebooters*:[1]

There is something more powerful than the brute force of bayonets: it is the idea whose time has come and hour struck.

That thought has itself evolved over the years, commonly being attributed to Aimard's more famous contemporary, Victor Hugo, who wrote in 1877 something similar in his *The History of a Crime*:[2]

One resists the invasion of armies; one does not resist the invasion of ideas.

Or, in the paraphrased form in which it is frequently quoted nowadays:

Nothing is stronger than an idea whose time has come.

The time has come to move from theory to practice in the fight against aging and favor of rejuveneering. It is our moral duty and our ethical responsibility to end the main cause of suffering in the world. The time has come to declare death to death.

All over the world, groups are emerging aware that this longed-for moment has arrived.[3] We have the technology then we have a moral duty. There are even emerging political parties whose explicit aim is to combat aging, as has already formally happened in Germany, the United States, and Russia. Activism should not be underestimated, nor activists, even if they are small groups. As the American anthropologist Margaret Mead said, it is precisely the conscious and committed individuals who transform humanity:[4]

Never doubt that a small group of thoughtful, committed citizens can change the world: indeed, it's the only thing that ever has.

We are reminded of another important historical reference when American President John Fitzgerald Kennedy launched his great challenge in 1961 to put a man on the Moon in just a decade. The challenge was huge, but the objective was met in 1969, although it even seemed impossible at the beginning. Let's take up another famous Kennedy phrase, but let's change the words "America" and "Americans" to "immortality" and "immoralists":

My fellow immoralists, ask not what "immortality" can do for you, ask what you can do for "immortality".

Although we repeat that the terms indefinite lifespan and indefinite life extension are more precise, everyone quickly grasps the idea of immortality (or, at least, the idea of "amortality"). Now we must consider these ideas to create a great world project against the common enemy of all humanity. Why not unite the whole planet in an Indefinite Youthfulness Project?

We need to have a comprehensive project that unites all of humanity based on the previous experience of previous successes such as the Manhattan Project, the Marshall Plan, the Apollo Program, the Human Genome Project, the International Space Station, the Human Brain Project, the International Thermonuclear Experimental Reactor, the CERN project, and so many other great multi-million dollar projects that have changed and continue to change the world.

We are witnessing the convergence of scientists, investors, large corporations and small startups working directly on issues of human aging and rejuvenation. We have science, we have money, and we have the ethical responsibility to end the major cause of human suffering. For the first time in history, we can do it, and we must do it. It is a historical imperative to achieve the first and greatest dream of all humanity.

We insist, at the risk of repeating ourselves, that we must not forget that every day, day after day, around the world, around 100,000 innocent people die from age-related diseases. One of the next may be you, or one of your loved ones. We can avoid it; we must avoid it. The sooner, the better. But we need your help, for it is everyone's fight against death. One alone cannot, all together we can.

The English evolutionary biologist JBS Haldane described the typical evolution of change processes, of great revolutions, starting in our minds:[5]

I suppose the process of acceptance will pass through the usual four stages:

1. This is worthless nonsense,
2. This is an interesting, but perverse, point of view,
3. This is true, but quite unimportant,
4. I always said so.

This is a revolution for your life, and my life, and the life of each one of us. We have in front of us a unique possibility and a great historical mission to fulfill. Given the dimension of this momentous project, the most severe mistake we might commit is to abandon the race before it begins. The opportunities for long and productive young life are far greater than the risks.

The future begins today. The future begins here. The future begins with us. The future begins with you today. Who if not you? When if not now? Where if not here? Join the revolution against aging and death! Death to death!

Notes

1. https://babel.hathitrust.org/cgi/pt?id=chi.087603619;view=1up;seq=67
2. https://www.amazon.com/History-Crime-Victor-Hugo/dp/384967696X
3. https://longevityalliance.org/history-of-the-international-longevity-alliance/
4. https://quoteinvestigator.com/2017/11/12/change-world/
5. https://quoteinvestigator.com/2017/06/13/acceptance/

Epilogue

All truth passes through three stages.
First, it is ridiculed.
Second, it is violently opposed.
Third, it is accepted as being self-evident.
Arthur Schopenhauer, 1819

I have known José Cordeiro for over a decade, and David Wood for a little less time, while our paths have crossed as speakers at multiple conferences related to aging and exponential technologies. The conference I founded to bring together scientists from all over the world, the Aging Research and Drug Discovery (ARDD), has become the largest forum in longevity biotechnology on the planet, which José and David have attended. Over time, my perspective on the field has changed. I once believed that we would be able to defeat aging in the very near future, but now I am more skeptical and pragmatic about the progress in longevity technology. I have gone from presenting optimistic outlooks at transhumanist conferences to taking a more conservative view and openly criticizing claims that I believe to be unrealistic. I now advocate for more research in cryobiology and other areas that may save the lives of millions of people who may not have access to clinically-validated rejuvenation interventions in the next few decades.

That is why I was surprised when José invited me to write the Epilogue to his book with David, *The Death of Death*, which includes very credible arguments and calls to action. I want to make it clear that the technological challenges that will help us defeat death are not yet solved, and will probably not be solved in the near future. My intention is not to discourage readers from supporting research in longevity, but rather to motivate them to actively engage in this field and to encourage others to do the same. The exaggerated hope that someone else will solve aging for us in the near future is false.

J. Cordeiro, D. Wood, *The Death of Death*, Copernicus Books,
https://doi.org/10.1007/978-3-031-28927-9

The reality is that most consumer-level longevity interventions explored by leading longevity clinics today focus on early disease diagnosis and lifestyle optimization, including diet, sleep, supplements, and basic pharmaceuticals intended to regulate metabolism. Even extreme lifestyle optimization will not dramatically increase lifespan. The most popular potential geroprotectors, like metformin and rapamycin, have not yet been tested in human clinical trials for longevity, and I believe that even if we do see life extension in healthy, athletic individuals, it will only be a few extra years. Furthermore, the idea that there is an abundance of novel therapeutics that are better than rapamycin in human clinical trials is not easy to sustain.

It typically costs over two billion dollars and takes over a decade to discover and develop a novel drug, with the process failing over 90% of the time. While we can use artificial intelligence to accelerate this process and improve the odds, we still need to conduct human clinical trials that usually take at least five years when testing the drug in a well-understood disease. And after the drug is approved for a disease, it may take a few years to prove that it slows down aging or repairs some of the aging-associated damage. I also want to remind the readers that gene therapy, cell therapy, and many other non-pharmacological interventions are also promising but still need more research and development before they can be clinically validated. It is important to support and engage in research in these areas, but it is also important to be realistic about the progress and challenges that lie ahead.

Making the situation worse was the deteriorating state of the global economy. Due to rapidly aging populations and excessive spending, some countries have accumulated massive debts that will be very hard to pay back. Inflation is eroding savings and is already hindering real economic growth. There is also a growing tension between the ultra-rich and the working class, which serves as a constant source of distraction. Instead of focusing the majority of their resources on aging research, countries are engaging in economic and even physical wars. When countries are fighting for survival, research in biotechnology is deprioritized. If this trend continues, many promising research groups will need to shift their focus from research to survival.

Nonetheless, there are also many reasons to be optimistic about the future. More and more scientists are switching to the longevity field from other disciplines. We are now seeing high school students choosing longevity biotechnology as their primary career path from the start. Most people who are not yet retired should consider a path of lifelong learning and career planning instead of planning for early retirement. With the new tools available to them in the near term, including AI and robotics, there will be many new alternative careers to pursue. You should not expose yourself to the risk of running out of savings after turning 100 years because you did not plan for it. Social security may not be there for you, and the ability to constantly acquire new skills and do useful work may help you survive and live a comfortable life.

It is important to stay focused on longevity and avoid getting easily distracted. Many successful politicians and other public figures have mastered the art of redirecting public attention to immediate, albeit minor, threats and engaging in blame games. It is easy to blame others for your shortcomings, and the easiest way to convince others is to first convince yourself. This is known as strategic self-deception. For example, as China's economy grows and gains a larger share of the global economy, some leaders in developed countries that are losing their dominance choose to rally their electorate against China and engage in destructive trade wars, instead of looking for opportunities to take advantage of the billions of people rising out of poverty and into prosperity. The right to life is the most fundamental human right, and in 2021, the average life expectancy in China was 2 years higher than in the USA. In the same year, the average Hong Kong citizen was expected to live, on average, 9 years longer than the average American. However, none of the major US politicians have prioritized longevity as part of their political agendas.

The next time you hear a politician making the argument that someone else is the cause of their problems, and providing justification for engaging in conflict, you may want to think twice before following that train of thought. Is it really worth it? Is it really worth wasting another chance to extend your life and the lives of other people on the planet? To achieve longevity, you need to truly want it, demand it, and fight for it. And considering the technological challenges that need to be overcome, as well as the costs and failure rates in the discovery of novel therapeutics, this quest requires massive global collaboration and integration.

The Death of Death highlights the importance of life and, while taking an optimistic view, provides a glimpse of hope for living substantially longer thanks to the many advances in science and technology. While I am more conservative in evaluating the overall progress in longevity biotechnology, as the companies I am building are now leading the field, and from that front seat, I can see the limits of the technological advances, it does not mean that I am ready to give up on the dream of significantly increasing longevity for everyone on the planet. In my opinion, this is the most important and impactful cause worth dedicating the remainder of my life to.

If you want to see the dream of longer lifespans, so colorfully painted in this international bestseller, realized within your lifetime, you will need to work hard to help make it happen. You should not expect others to do the work for you. Ideally, you should consider contributing to the field intellectually, getting deeper into the field, and inventing new methods for life extension. In the case that you are not capable of doing this, you should consider bringing more resources into longevity biotechnology and motivating others to join. And if you cannot do that, you can still be helpful. Even efforts to restore global peace, stop counterproductive trade wars, increase collaboration between countries and organizations, and help avoid a possible economic collapse, will help researchers keep fighting for your life and the countless lives of other humans - lives that would otherwise be lost to aging and death.

I do not disagree with José and David since there is an important chance that we will see dramatic increases in longevity within our lifetimes. I just think that it is not guaranteed, it might take much longer, and we need to work much harder. Therefore, there is no time to waste. As the famous quote says:

If not us, who?

And if not now, when?

Alex Zhavoronkov, PhD
Founder and CEO, Insilico Medicine, and author of *The Ageless Generation*

Appendix: Big Chronology of Life on Earth

> Two possibilities exist:
> Either we are alone in the universe or we are not,
> both are equally terrifying.
> **Sir Arthur C. Clarke, 1962**

> Live long and prosper.
> yIn nI' yISIQ 'ej yIchep (Klingon pronunciation).
> dif-tor heh smusma (Vulcan pronunciation).
> **Commander Spock from Vulcan in the Spaceship USS Enterprise, 2260**

To put into perspective a complete chronology and evolution of life on our tiny planet Earth, we summarize here what we consider the most relevant information from the very distant past to our immediate future. The objective is to reach a better understanding about the long-term evolution of life, considering also the nature of exponential changes.

Big History is a new discipline that allows us to analyze with a multidisciplinary focus the way events follow each other throughout time. Starting with a huge time scale from the faraway past to the present, we can see that there is an acceleration of the speed of changes, that should continue now thanks to exponential technologies. The great futurist Ray Kurzweil, in his best-seller *The Singularity is Near*, explains very well the acceleration of these changes, and that is why we use some of his predictions to the end of the twenty-first century.

Interested readers are invited to contact us directly to continue making this chronology better in the future. All comments are more than welcome by contacting the authors.

© The Author(s), under exclusive license to Springer Nature Switzerland AG 2023
J. Cordeiro, D. Wood, *The Death of Death*, Copernicus Books,
https://doi.org/10.1007/978-3-031-28927-9

Millions of Years Ago (Ma)

~13,800 Ma	Big Bang and formation of the known Universe
~12,500 Ma	Milky Way Galaxy formation
~4600 Ma	Solar System formation
~4500 Ma	Earth formation
~4300 Ma	First water concentration on Earth
~ 4000 Ma	First unicellular life (prokaryotes without cellular nucleus)
~ 4000 Ma	LUCA, our Last Universal Common Ancestor, was born
~ 3500 Ma	Oxygen concentration rises on Earth atmosphere
~ 3000 Ma	First photosynthesis in simple unicellular organisms
~ 2000 Ma	Evolution of unicellular prokaryotes (without nucleus) into eukaryotes (with nucleus)
~ 1500 Ma	First multicellular eukaryote organisms
~ 1200 Ma	First sexual reproduction (germinal and somatic cells appear)
~ 600 Ma	First invertebrate marine animals
~ 540 Ma	Cambrian explosion and appearance of multiple species
~ 520 Ma	First vertebrate marine animals
~ 440 Ma	Evolution from marine life to terrestrial life (first plants on dry land)
~ 360 Ma	First terrestrial plants with seeds, and first crabs
~ 300 Ma	First reptiles
~ 250 Ma	First dinosaurs
~ 200 Ma	First mammals, and first birds
~ 130 Ma	First angiosperm plants (with flowers)
~ 65 Ma	Extinction of dinosaurs and development of primates
~ 15 Ma	*Hominidae* family (big primates) appears
~ 3.5 Ma	First tools made of stone
~ 2.5 Ma	*Homo* genus appears
~ 1.5 Ma	First use of fire
~ 0.8 Ma	First time cooking was used
~ 0.5 Ma	First time clothes were used
~ 0.2 Ma	*Homo sapiens* species appears
~ 0.1 Ma	*Homo sapiens sapiens* comes out of Africa and starts colonizing planet Earth

Thousands of Years Ago

~ 40,000 BC	Rock paintings appear, symbols of deities, fertility and death
~ 20,000 BC	Lighter skin evolution due to migration to regions with less solar exposure
~ 5000 BC	Neolithic proto-writing appears
~ 4000 BC	Possible invention of the wheel in Mesopotamia

~ 3500 BC   Egyptians invent hieroglyphs and Sumerians cuneiform writing
~ 3300 BC   Documented use of herbology and physiotherapy in China and Egypt
~ 3000 BC   Papyrus is invented in Egypt and clay tablets in Mesopotamia
~ 2800 BC   Chinese emperor Shennong compiles a text with acupuncture techniques
~ 2600 BC   Imhotep, priest and doctor, is considered the God of Medicine in Egypt
~ 2500 BC   Documented use of Ayurveda medicine in India
~ 2000 BC   The *Code of Hammurabi* establishes rules to exercise medicine in Babylon
~ 650 BC   Assurbanipal compiles 800 tablets about medicine in the library of Nineveh
~ 450 BC   Xenophanes of Colophon examines fossils and speculates about the evolution of life
420 BC   Hippocrates writes the *Hippocratic Treaties* and creates the *Hippocratic oath*
350 BC   Aristotle writes about evolutionary biology and tries to classify animals
300 BC   Herophilos of Chalcedon makes medical dissections on humans
100 BC   Asclepiades of Bithynia imports Greek medicine to Rome and funds the Methodic School

First Millennium AD

180 AD   Greek doctor Galen of Pergamon studies the connection between paralysis and the spinal cord
219 AD   Zhang Zhongjing publishes the *Shanghan Lun* (*Treatise on Cold Damage Disorders*) in China
250 AD   Foundation of a school of tribal medicine in Monte Alban, Mexico
390 AD   Oribasius of Pergamon compiles the *Medical Collections in* Constantinople
400 AD   First Christian hospital founded by Saint Fabiola in Rome
630 AD   Isidore of Seville compiles his great work *The Etymologies*
870 AD   Persian doctor Ali ibn Sahl Rabban al-Tabari writes a medical encyclopedia in Arab
910 AD   Persian doctor Rasis identifies the difference between smallpox and measles

1000–1799 AD

1030   Persian polymath Avicenna writes the Cannon of Medicine that would be used until eighteenth century
1204   Pope Innocent III organizes the first Holy Spirit hospital in Rome

1403	Quarantine against the Black Death pandemic in Venice (after already killing million in Europe)
1541	Swiss doctor Paracelsus made great progress in medicine (surgery and toxicology)
1553	Spanish doctor Miguel Servet studies pulmonary circulation (and is burnt at the stake for heresy)
1590	Microscope is invented in the Netherlands and makes medicine move forward faster
1665	English scientist Robert Hooke uses the microscope to identify cells (and popularizes that name)
1675	Dutch scientist Anton van Leeuwenhoek starts microbiology with microscopes
1774	English scientist Joseph Priestley discovers oxygen and starts modern chemistry
1780	US polymath Benjamin Franklin writes about curing aging and human preservation
1796	English doctor Edward Jenner develops the first effective vaccine against smallpox
1798	English scholar Thomas Malthus argues about food production and human overpopulation

1800–1899 AD

1804	Global population reaches 1,000,000,000 people
1804	French doctor René Laennec invents the stethoscope
1809	French scientist Jean-Baptiste Lamarck proposes the first theory of evolution
1818	English doctor James Blundell performs the first successful blood transfusion
1828	German scientist Christian Ehrenberg coins the word bacterium ("cane" in Greek)
1842	US doctor Crawford Long accomplishes the first surgery with anesthesia
1858	German doctor Rudolf Virchow publishes his cell theory
1859	English scientist Charles Darwin publishes *The origin of species* in London
1865	Austrian monk Gregor Mendel discovers the laws of genetics
1869	Swiss doctor Friedrich Miescher discovers DNA for the first time
1870	Scientists Louis Pasteur and Robert Koch publish the microbial theory of infections
1882	French scientist Louis Pasteur develops a vaccine against rabies

1890	Walter Flemming and others describe the chromosome distribution during cellular division
1892	German biologist August Weismann proposes the "immortality" of germ cells
1895	German physicist Wilhelm Conrad Röntgen discovers X rays and their medical uses
1896	French physicist Antoine Henri Becquerel discovers radioactivity
1898	Dutch scientist Martinus Beijerinck discovers the first virus and starts virology

1900–1959 AD

1905	English biologist William Bateson coins the term "genetics"
1906	English scientist Frederick Hopkins describes vitamins and associated illnesses
1906	German doctor Alois Alzheimer describes the disease named after him
1906	Spanish scientist Santiago Ramón y Cajal receives the Nobel Prize for his studies about the nervous system
1911	Thomas Hunt Morgan demonstrates that genes reside in chromosomes
1922	Russian scientist Aleksandr Oparin proposes a theory about the origin of life on Earth
1925	French biologist Edouard Chatton coins the words prokaryote and eukaryote
1927	Global population reaches 2,000,000,000 people
1927	First vaccines against tetanus and tuberculosis
1928	English scientist Alexander Fleming discovers penicillin (first antibiotic)
1933	Polish scientist Tadeus Reichstein synthesizes a vitamin for the first time (vitamin C, ascorbic acid)
1934	Scientists working at Cornell University discover caloric restriction for life extension in mice
1938	A coelacanth (considered a "living fossil") was fished in the south of Africa
1950	First synthetic antibiotic is developed
1951	Artificial insemination of cattle starts with cryopreserved semen
1951	HeLa (Henrietta Lacks) cancer cells are discovered to be "biologically immortal"
1952	US doctor Jonas Salk develops a vaccine against poliomyelitis
1952	US chemist Stanley Miller conducts experiments about the origin of life
1952	First cloning experiments with frog eggs
1953	Scientists James D. Watson and Francis Crick demonstrate DNA's double helix structure
1954	US doctor Joseph Murray transplants the first human kidney
1958	US doctor Jack Steele coins the word "bionic"

1959   Global population reaches 3,000,000,000 people
1959   Spanish scientist Severo Ochoa receives the Nobel Prize for his work about DNA and RNA

1960–1999 AD

1961   Spanish biochemist Joan Oró advances his theories about the origin of life
1961   US scientist Leonard Hayflick discovers a limit on cellular division
1967   US academic James Bedford becomes the first patient in cryopreservation
1967   South African doctor Christiaan Barnard makes the first human heart transplant
1972   Discovery that the DNA composition in humans and gorillas is almost 99% similar
1974   Global population reaches 4,000,000,000 people
1975   Different scientists finally discover the structure of telomeres (first considered in 1933)
1978   First human being is born thanks to artificial insemination (Louise Brown in England)
1978   Stem cells discovered in the blood of an umbilical cord
1980   World Health Organization declares smallpox officially eradicated worldwide
1981   First stem cells (from mice) developed "in vitro"
1982   Humulin (drug for diabetes) is the first biotech product approved by the FDA
1985   Australian-American biologist Elizabeth Blackburn identifies the telomerase enzyme
1986   HIV (Human Immunodeficiency Virus) is identified as the cause of AIDS
1987   Global population reaches 5,000,000,000 people
1990   Human Genome Project starts as a great effort lead by several governments
1990   First gene therapy is approved to treat an immune disorder
1990   FDA approves the first genetically modified organism (Flavr Savr tomato)
1993   US biologist Cynthia Kenyon increases several times the lifespan of *C. elegans*
1995   US scientist Caleb Finch describes negligible senescence in some animals
1996   Scottish scientist Ian Wilmut clones Dolly, the first cloned mammal (a sheep)
1998   First embryonic stem cells isolated in young human embryos
1999   Global population reaches 6,000,000,000 people

2000–2022 AD

2001	US scientist Craig Venter announces his sequence of the human genome (based on his own DNA)
2002	First artificial virus (polio virus) is completely created by scientists
2003	Human Genome Project ends officially, with both public and private participation and projects
2003	English scientist Aubrey de Grey and his colleagues create the Methuselah Foundation
2004	SARS epidemic is contained a year after its start (genome sequenced in months)
2006	Japanese scientist Shinya Yamanaka generates induced pluripotent stem cells in Kyoto
2008	Spanish biologist María Blasco announces the life extension of mice at CNIO in Madrid
2009	English scientist Aubrey de Grey and his colleagues create the SENS Research Foundation
2009	Nobel Prize on Physiology and Medicine for studies on telomeres and telomerase
2010s	First Bridge towards indefinite lifespans using current technologies (Ray Kurzweil)
2010	US scientist Craig Venter announces the creation of the first artificial bacterium (Synthia)
2010	Nobel Prize on Physiology and Medicine for the development of in vitro fertilization
2011	Global population reaches 7,000,000,000 people
2011	French researches achieve the rejuvenation of human cells "in vitro"
2012	Nobel Prize on Physiology and Medicine for cloning and cell reprogramming (pluripotent cells)
2013	First rat kidney produced "in vitro" in the USA
2013	First human liver produced with stem cells in Japan
2013	Google announces the creation of Calico (California Life Company) to cure aging
2014	IBM expands the use of its intelligent medical system called Doctor Watson
2014	Korean-American doctor Joon Yun creates the Palo Alto Longevity Prize
2015	First vaccine against the virus of Ebola hemorrhagic fever
2016	Facebook chairman Mark Zuckerberg announces that it will be possible to cure "all diseases"
2016	Microsoft scientists announce that they should be able to cure cancer within 10 years
2016	German entrepreneur Michael Greve founds the Forever Healthy Foundation

2017	Spanish scientist Juan Carlos Izpisúa announces that his Salk Institute team has been able to rejuvenate mice 40%
2018	First commercial treatment with gene therapy using CRISPR
2018	Birth of first CRISPR babies to avoid HIV infections in China
2019	FDA approval of the first senolytic treatments for life extension
2019	Nature magazine carries the report "First hint that body's 'biological age' can be reversed" featuring the TRIIM trial led by Greg Fahy
2020	Covid2019 virus genome is sequenced in weeks and new mRNA vaccines are developed in days
2020	AlphaFold (AI developed by Google's DeepMind) solves the protein folding in biology
2020	Nobel Prize on Physiology and Medicine for CRISPR research
2020	Eyes of blind mice rejuvenated with epigenetic reprogramming by Australian biologist David Sinclair at Harvard University
2021	American Jeff Bezos, Russian Yuri Milner and other billionaires create Altos Labs to reverse the human aging process
2021	First transplant of a pig's kidney into a brain-dead human showed that xenotransplants can work
2021	Largest vaccination campaign in history to contain the Covid19 pandemic after the production of more than 9 billion vaccines in 1 year
2022	First transplant of a pig's heart into a human showed that xenotransplants worked
2022	FDA approves first experimental treatments for hemophilia and Alzheimer's disease
2022	Saudi Arabia announces the creation of Hevolution (Health+Evolution) Foundation to finance billions into aging research
2022	English scientist Aubrey de Grey and his colleagues create the LEV (Longevity Escape Velocity) Foundation
2022	Global population reaches 8,000,000,000 people

2023 AD–2029 AD (Some Possibilities)

2023	First experimental clinical trials with mRNA vaccines for cancer
2024	First experimental clinical trials with mRNA vaccines for malaria and HIV
2025	Robust Mouse Rejuvenation is achieved in trials sponsored by LEV Foundation
2025	Molecular assemblers (nanotechnology) become possible (Ray Kurzweil)

2020s	Second Bridge towards indefinite lifespans using biotechnology (Ray Kurzweil)
2020s	Human clinical trials for longevity with metformin and rapamycin
2020s	Worldwide eradication of poliomyelitis
2020s	Worldwide eradication of measles
2020s	Successful vaccines approved against malaria and HIV
2020s	Cure for the majority of cancers
2020s	Cure for Parkinson's disease
2020s	3D bioprinting of simple human organs
2020s	Commercial cloning of human organs with own cells from patients
2020s	Beginning of commercial rejuvenation treatments with stem cells and telomerase
2020s	AI and robot doctors complement and supplement human doctors
2020s	Telemedicine spreads worldwide
2020s	First manned trips to Mars (Elon Musk)
2029	Longevity Escape Velocity (LEV) or the "Methuselarity" is reached (Ray Kurzweil)
2029	An advanced AI finally passes Alan Turing's test (Ray Kurzweil)

After 2030 AD (More Possibilities)

2030s	Third Bridge towards indefinite lifespans using nanotechnology (Ray Kurzweil)
2030s	Cure for Alzheimer's disease
2030s	Worldwide eradication of malaria
2030s	Worldwide eradication of HIV
2030s	Consolidation of the first human colony in Mars (Elon Musk)
2037	Global population reaches 9,000,000,000 people
2039	Mental transfer from brain to brain becomes possible (Ray Kurzweil)
2040s	Final Bridge towards indefinite lifespans and immortality using AI (Ray Kurzweil)
2040s	Interplanetary Internet connects to Earth, Moon, Mars and spaceships
2045	Aging is cured and death becomes optional (Ray Kurzweil)
2045	The Singularity: AI surpasses all human intelligence (Ray Kurzweil)
2049	Distinction between reality and virtual reality disappears (Ray Kurzweil)
2050	Humanoid robots win English football cup (British Telecom)
2050s	First reanimations of cryopreserved patients (Ray Kurzweil)
2072	Picotechnology starts (pico is one thousand times smaller than nano, Ray Kurzweil)
2099	Femtotechnology starts (femto one thousand times smaller than pico, Ray Kurzweil)
2099	Lifespan becomes irrelevant in a world of "amortality"

Acknowledgments

> Gratitude is the sign of noble souls.
> **Aesop, fifth century BC**

> If I have seen further, it is by standing upon the shoulders of giants.
> **Isaac Newton, 1676**

> Victory has a hundred fathers, but defeat is an orphan.
> **John Fitzgerald Kennedy, 1961**

This is a book about life, for life, and to life. The first acknowledgments are to our families, who have allowed us to get here. Though, even more so than to our direct families, our gratitude is also for our first hominid ancestors in Africa millions of years ago, and long before that the first unicellular organisms from which all life descends on our small planet.

Many thanks to our colleagues and friends at universities and institutions such as the Universidad Autónoma of Madrid, Barcelona, Birmingham, Berkeley, Cambridge, University College London, Complutense, Georgetown, Harvard, Higher School of Economics, INSEAD, King's College, Kyoto, Liverpool, MIPT, MIT, Oxford, Politécnica de Madrid, Singularity, Seoul, Singapore, Sophia, Stanford, Tecnológico de Monterrey, Tokyo, Waseda, Westminster, Yonsei, and many others in different parts of the world. We would also like to thank people in other visionary organizations such as AfroLongevity, Alcor Life Extension Foundation, Alliance for Longevity Initiatives, American Academy of Anti-Aging Medicine, Club of Rome, Coalition for Radical Life Extension, Cryonics Institute, European Biostasis

© The Author(s), under exclusive license to Springer Nature Switzerland AG 2023
J. Cordeiro, D. Wood, *The Death of Death*, Copernicus Books,
https://doi.org/10.1007/978-3-031-28927-9

Foundation, FitTech Summit, Forever Healthy Foundation, International Longevity Alliance, KrioRus, LEV (Longevity Escape Velocity) Foundation, Life Extension Foundation, Lifeboat Foundation, London Futurists, Madrid Singularity, Methuselah Foundation, HumanityPlus, SENS Research Foundation, SingularityNET, Southern Cryonics, The Millennium Project, TechCast Global, Tomorrow Biostasis, TransHumanCoin, Transhumanist Party, VitaDAO, World Academy of Art and Science, World Future Society, World Futures Studies Federation, and other futurist groups around the world.

Individually, we would like to express our gratitude to the scientists, researchers, investors, communicators, advocates, economists, and politicians who are actively working toward the radical extension of life, including, alphabetically by last name: Johnny Adams, Mark Allen, Bruce Ames, Omri Amirav-Drory, Bill Andrews, Christian Angermayer, Sonia Arrison, John Asher, Anthony Atala, Jacques Attali, Peter Attia, Steven Austad, Charles Awuzie, Mustafa Aykut, Rafael Badziag, Ronald Bailey, Ben Ballweg, Joe Bardin, Hal Barron, Nir Barzilai, Nir Barzilai, Kate Batz, Boris Bauke, Alexandra Bause, Andrea Bauer, Eckhart Beatty, Heiner Benking, Joanna Bensz, Adriane Berg, Marc Bernegger, Ben Best, Jeff Bezos, Santiago Bilinkis, Evelyne Bischof, Hans Bishop, Marko Bitenc, Victor Bjork, Celia Black, Gil Blander, María Blasco, Gunter Boden, Felix Bopp, Nick Bostrom, Alon Braun, Nicklas Brendborg, Charles Brenner, Sergey Brin, Jan Bruch, Sebastian Brunemeier, Martha Bucaram, Sven Bulterijs, Patrick Burgermeister, Per Bylund, Ismael Cala, Judith Campisi, Hector Casanueva, Sophie Chabloz, Calum Chace, Al Chalabi, Puruesh Chaudhary, Matthew Chavira, Nathan Cheng, Nicolas Chernavsky, Pedro Chomnalez, Epaminondas Christophilopoulos, George Church, Zina Cinker, Günter Clar, Vitto Claut, Sven Clemann, James Clement, Didier Coeurnelle, Margaretta Colangelo, Kristin Comella, Keith Comito, Irina Conboy, Nichola Conlon, Franco Cortese, Kat Cotter, Glenn Cripe, Walter Crompton, Shermon Cruz, Attila Csordas, Adrian Cull, Cornelia Daheim, Stephanie Dainow, Nikola Danaylov, Stanley Dao, Rafael de Cabo, João Pedro de Magalhães, Peter de Keizer, Heitor Gurgulino de Souza, Yuri Deigin, Brian Delaney, Dinorah Delfin, Marco Demaria, Laura Deming, Jyothi Devakumar, Bobby Dhadwar, Peter Diamandis, Mara di Berardo, Eric Drexler, Allison Duettmann, David Ewing Duncan, George Dvorsky, Victor J. Dzau, Anastasia Egorova, Dan Elton, Nick Engerer, María Entraigues-Abramson, Collin Ewald, Lisa Fabiny-Kiser, Gregory Fahy, Bill Faloon, Peter Fedichev, Ruben Figueres, Zan Fleming, Kristen Fortney, Michael Fossel, Thomas Frey, Robert Freitas, Petr Fridrich, Patri Friedman, Garry Jacobs, Steven Garan, Eleanor Garth, Maximilian Gaub, Titus Gebel, Michael Geer, Alan Gehrich, Anastasiya Giarletta,

Sebastian Giwa, Vadim Gladyshev, Jerome Glenn, David Gobel, Ben Goertzel, Tyler Golato, Robert Goldman, Vera Gorbunova, Ted Gordon, Rodolfo Goya, Michael Greve, Ivana Greguric, Adam Gries, Greg Grinberg, Magdalena Groselj, Dan Grossman, Terry Grossman, Leonard Guarente, Bill Halal, Ian Hale, Mark Hamalainen, David Hanson, William Haseltine, Petra Hauser, Lou Hawthorne, Kenneth Hayworth, Wei-Wu He, Jean Hébert, Andrew Hessel, Steve Hill, Rudi Hoffman, Steve Horvath, Matthias Hornberger, Ted Howard, Edward Hudgins, Gary Hudson, Barry Hughes, Reyhan Huseynova, Paul Hynek, Generoso Ianniciello, Cairn Idun, Tom Ingoglia, Niccolò Invidia, Laurence Ion, Anca Iovita, Javier Irizarry, Salim Ismail, Zoltan Istvan, Juan Carlos Izpisúa Belmonte, Garry Jacobs, Naveen Jain, Ravi Jain, Sumit Jamuar, Ana Jerkovic, Bryan Johnson, Tanya Jones, Matt Kaeberlein, Michio Kaku, Osinakachi Akuma Kalu, Charlie Kam, Dmitry Kaminskiy, Bill Kapp, Natalia Karbasova, David Karow, Alexander Karran, Steve Katz, Sandra Kaufmann, Peter Kaznacheev, Emil Kendziorra, Brian Kennedy, Magomed Khaidakov, Daria Khaltourina, Faraz Khan, Mehmood Khan, Aaron King, James Kirkland, Ronald Klatz, Richard Klausner, Eric Klien, Randal Koene, Michael Kope, Daniel Kraft, Guido Kroemer, Anton Kulaga, Ray Kurzweil, Marios Kyriazis, Yosi Lahad, James Lark, Alessandro Lattuada, Gordan Lauc, Nikolina Lauc, Newton Lee, Eugen Leitl, Jean-Marc Lemaitre, Christine Lemstra, Gerd Leonhard, Kate Levchuk, Michael Levin, Morgan Levine, Caitlin Lewis, John Lewis, Martin Lipovšek, Dylan Livingston, Scott Livingston, Bruce Lloyd, Valter Longo, Carlos López-Otín, Miguel López de Silanes, Epi Ludvik, Michael Lustgarten, Robert Konrad Maciejewski, Dip Maharaj, Andrea B. Maier, Polina Mamoshina, Keith Mansfield, Dana Marduk, Milan Maric, Juan Martínez-Barea, Eric Martinot, Nuno Martins, Max Marty, Robert Luke Mason, Stephen Matlin, John Mauldin, Raymond McCauley, Danila Medvedev, Oliver Medvedik, Jim Mellon, Jason Mercurio, Ralph Merkle, Bertalan Meskó, Jamie Metzl, Phil Micans, Fiona Miller, Kai Micah Mills, Elena Milova, Chris Mirabile, Barun Mitra, Eli Mohamad, Kelsey Moody, Max More, Alexey Moskalev, Wolfgang Müller, Elon Musk, Ronjon Nag, Torsten Nahm, Brent Nally, José Navarro-Betancourt, Phil Newman, Pat Nicklin, Suresh Nirody, Patrick Noack, Guido Núñez-Mujica, Matthew O'Connor, Andrea Oddone, Martin O'Dea, Ines O'Donovan, Ryan O'Shea, Alejandro Ocampo, Concepción Olavarrieta, Jay Olshansky, David Orban, Dean Ornish, Erik Ferdinand Øverland, Larry Page, Cesar Paiva, Francisco Palao, Liz Parrish, Linda Partridge, Ira Pastor, David Pearce, Kevin Perrott, Michael Perry, Steve Perry, Leon Peshkin, Christine Peterson, James Peyer, Maximus Peto, Miri Polachek, Mila Popovich, Frances Pordes, Alexander Potapov, Ronald A. Primas, Giulio Prisco, Marco Quarta, Ana Quintero,

224   Acknowledgments

Michael Rae, Carrie Radomski, Brenda Ramokopelwa, Thomas Rando, Ashish Rajput, Reason, Jovan David Rebolledo, Antonio Regalado, Tobias Reichmuth, Robert J.S. Reis, Denisa Rensen, Michael Ringel, Ramón Risco, Eric Risser, Tony Robbins, Pascal Rode, Edwina Rogers, Michael Rose, Tom Ross, Maurizio Rossi, Gabriel Rothblatt, Martine Rothblatt, Avi Roy, Danielle Ruiz, Sergio Ruiz, Mary Ruwart, Mark Sackler, Paul Saffo, Roberto Saint-Malo, Anders Sandberg, Jelena Sarenac, Morten Scheibye-Knudsen, Boris Schmalz, Matthew Scholz, Ken Schoolland, Frank Schüler, Kurt Schuler, Björn Schumacher, Andrew J. Scott, Kenneth Scott, Tony Seba, Vittorio Sebastiano, Elena Segal, Thomas Seoh, Manuel Serrano, Yair Sharan, Jin-Xiong She, Eleanor Sheekey, Larisa Sheloukhova, Lori L. Shemek, David Shumaker, Skip Sidiqi, Bernard Siegel, Felipe Sierra, Michal Siewierski, Jason Silva, David A. Sinclair, Richard Siow, Hannes Sjöblad, Mark Skousen, John Smart, Jacek Spendel, Paul Spiegel, Petr Sramek, Ilia Stambler, Brad Stanfield, Andrew Steele, Clemens Steinek, Gregory Stock, Gennady Stolyarov II, Alexandra Stolzing, Jim Strole, Danny Sullivan, Csaba Szabo, Rohit Talwar, Ufuk Tarhan, Emma Teeling, Oleg Teterin, Peter Thiel, Mohan Tikku, Mariana Todorova, Marco Tricomi, Peter Tsolakides, Alexey Turchin, Roey Tzezana, Maximilian Unfried, Iruña Urruticoechea, Arin Vahanian, Neal VanDeRee, Yossi Vardi, Álvaro Vargas-Llosa, Zack Varkaris, Harold Varmus, Kyle Varner, Craig Venter, Kris Verburgh, Eric Verdin, Natasha Vita-More, Sanja Vlahovic, Peter Voss, Chip Walter, Kevin Warwick, Simon Waslander, Amy Webb, Michael West, Todd White, Kristen Willeumier, Robert Wolcott, Tina Wood, Peter Xing, Shinya Yamanaka, Sergey Young, Peter Zemsky, Alex Zhavoronkov, Mikolaj Zielinski, Oliver Zolman, and Ibon Zugasti, among many others inspiring with their visionary ideas and work.

Finally, we would like to thank all the readers of this book for their interest, and kindly ask them to write us with ideas, suggestions, corrections, or any additional comments to the contact address of the book's website. All messages are more than welcome to continue improving the book, of which future editions will also include in the acknowledgments' section the names of the people who write to us. Your comments will allow this work to reach new readers and make ideas more precise, please. Your recommendation of this book to others is equally important to help with scientific advances.

This is a perfectible book, like life itself. It is also an "immortal" book that will continue to evolve and change, like "immortal" life in the future. This is a continuous improvement work, thanks to readers like you.

All your suggestions are welcome, forever and ever!

Bibliography

One glance at a book and you hear the voice of another person, perhaps someone dead for 1,000 years. To read is to voyage through time... Books break the shackles of time – proof that humans can work magic.
Carl Sagan, 1980

How many a man has dated a new era in his life from the reading of a book.
Henry David Thoreau, 1854

1. Alberts, Bruce. (2014). *Molecular Biology of the Cell. 6th Edition.* Garland Science.
2. Alexander, Brian. (2004). *Rapture: A Raucous Tour of Cloning, Transhumanism, and the New Era of Immortality.* Basic Books.
3. Alexandre, Laurent. (2011). *La mort de la mort: Comment la technomédicine va bouleverser l'humanité.* Editions Jean-Claude Lattès.
4. Alighieri, Dante. (2008 [1321]). *The Divine Comedy.* Chartwell Books.
5. Andrews, Bill & Cornell, Jon. (2017). *Telomere Lengthening: Curing All Disease Including Aging and Cancer.* Sierra Sciences.
6. Andrews, Bill & Cornell, Jon. (2014). *Curing Aging: Bill Andrews on Telomere Basics.* Sierra Sciences.
7. Arking, Robert. (2006). *The Biology of Aging: Observations and Principles.* Oxford University Press.
8. Arrison, Sonia. (2011). *100 Plus: How the Coming Age of Longevity Will Change Everything, From Careers and Relationships to Family and Faith.* Basic Books.
9. Asimov, Isaac. (1993). *Asimov's New Guide to Science.* Penguin Books Limited.
10. Austad, Steven N. (2022). *Methuselah's Zoo: What Nature Can Teach Us about Living Longer, Healthier Lives.* The MIT Press.
11. Austad, Steven N. (1997). *Why We Age: What Science Is Discovering About the Body's Journey Through Life.* John Wiley & Sons, Inc.
12. Bailey, Ronald. (2005). *Liberation Biology: The Scientific and Moral Case for the Biotech Revolution.* Prometheus Books.

© The Author(s), under exclusive license to Springer Nature Switzerland AG 2023
J. Cordeiro, D. Wood, *The Death of Death*, Copernicus Books,
https://doi.org/10.1007/978-3-031-28927-9

13. Barzilai, Nir. (2020). *Age Later: Health Span, Life Span, and the New Science of Longevity*. St. Martin's Press.
14. BBVA, OpenMind. (2017). *The Next Step: Exponential Life*. BBVA, OpenMind.
15. Becker, Ernest. (1973). *The Denial of Death*. Free Press.
16. Blackburn, Elizabeth & Epel, Elissa. (2018). *The Telomere Effect: A Revolutionary Approach to Living Younger, Healthier, Longer*. Grand Central Publishing.
17. Blasco, María & Salomone, Mónica G. (2016). *Morir joven, a los 140: El papel de los telómeros en el envejecimiento y la historia de cómo trabajan los científicos para conseguir que vivamos más y mejor*. Paidós.
18. Bostrom, Nick. (2005). "A History of Transhumanist Thought." *Journal of Evolution and Technology*, Vol. 14 Issue 1, April 2005.
19. Bova, Ben. (1998). Immortality: How Science is Extending Your Life Span, and Changing the World. Avon Books.
20. Brendborg, Nicklas. (2023). Jellyfish Age Backwards: Nature's Secrets to Longevity. Little, Brown and Company.
21. Bulterijs, Sven; Hull, Raphaella S.; Bjork, Victor C. & Roy, Avi G. (2015). "It is time to classify biological aging as a disease." Frontiers in Genetics 6:205.
22. Carlson, Robert H. (2010). Biology is Technology: The promise, peril, and new business of engineering life. Harvard University Press.
23. Cave, Stephen. (2012). Immortality: The Quest to Live Forever and How It Drives Civilization. Crown.
24. Chaisson, Eric. (2005). Epic of Evolution: Seven Ages of the Cosmos. Columbia University Press.
25. Church, George M. and Regis, Ed. (2012). Regenesis: How Synthetic Biology will Reinvent Nature and Ourselves. Basic Books.
26. Clarke, Arthur C. (1984 [1962]). Profiles of the Future: An Inquiry into the Limits of the Possible. Henry Holt and Company.
27. Clement, James W. (2021). The Switch. Gallery Books.
28. Comfort, Alex. (1964). Aging: The Biology of Senescence. Routledge & Kegan Paul.
29. Condorcet, Marie-Jean-Antoine-Nicolas de Caritat. (1979 [1795]). Sketch for a Historical Picture of the Progress of the Human Mind. Greenwood Press.
30. Cordeiro, José (ed.). (2014). Latinoamérica 2030: Estudio Delphi y Escenarios. Lola Books.
31. Cordeiro, José. (2010). Telephones and Economic Development: A Worldwide Long-Term Comparison. Lambert Academic Publishing.
32. Cordeiro, José. (2007). El Desafío Latinoamericano... y sus Cinco Grandes Retos. McGraw-Hill Interamericana.
33. Coeurnelle, Didier. (2013). Et si on arrêtait de vieillir! FYP éditions.
34. Critser, Greg. (2010). Eternity Soup: Inside the Quest to End Aging. Crown.
35. Danaylov, Nikola. (2016). Conversations with the Future: 21 Visions for the 21st Century. Singularity Media, Inc.

36. Darwin, Charles. (2003 [1859]). The Origin of the Species. Fine Creative Media.
37. Dawkins, Richard. (1976). The Selfish Gene. Oxford University Press.
38. De Grey, Aubrey & Rae, Michael. (2008). Ending Aging: The Rejuvenation Breakthroughs That Could Reverse Human Aging in Our Lifetime. St. Martin's Press.
39. De Grey, Aubrey; Ames, Bruce N.; Andersen, Julie K.; Bartke, Andrzej; Campisi, Judith; Heward, Christopher B.; McCarter, Roger JM & Stock, Gregory. (2002). "Time to talk SENS: critiquing the immutability of human aging." Annals of the New York Academy of Sciences. Vol. 959; pp. 452–462.
40. De Grey, Aubrey. (1999). The mitochondrial free radical theory of aging. Landes Bioscience.
41. De Magalhães, João Pedro, Curado, J. & Church, George M. (2009). "Meta-analysis of age-related gene expression profiles identifies common signatures of aging." Bioinformatics, 25(7), pp. 875–881.
42. Deep Knowledge Ventures. (2018). AI for Drug Discovery, Biomarker Development and Advanced R&D. Deep Knowledge Ventures.
43. DeLong, J. Brad. (2000). "Cornucopia: The Pace of Economic Growth in the Twentieth Century." NBER Working Papers 7602.
44. Diamandis, Peter H. & Kotler, Steven. (2016). Bold: How to Go Big, Create Wealth and Impact the World. Simon & Schuster.
45. Diamandis, Peter H. & Kotler, Steven. (2012). Abundance: The Future is Better Than You Think. Free Press.
46. Diamond, Jared M. (1997). Guns, Germs, and Steel: The Fates of Human Societies. W. W. Norton & Co.
47. Drexler, K. Eric. (2013). Radical Abundance: How a Revolution in Nanotechnology Will Change Civilization. PublicAffairs.
48. Drexler, K. Eric. (1987). Engines of Creation: The Coming Age of Nanotechnology. Anchor Books.
49. Dyson, Freeman J. (2004 [1984]): Infinite in All Directions. Harper Perennial.
50. Ehrlich, Paul. (1968). The Population Bomb. Sierra Club/Ballantine Books.
51. Emsley, John. (2011). Nature's Building Blocks: An A-Z Guide to the Elements. Oxford University Press.
52. Ettinger, Robert. (1972). Man into Superman. St. Martin's Press.
53. Ettinger, Robert. (1964). The Prospect of Immortality. Doubleday.
54. Fahy, Gregory et al. (ed.). (2010). The Future of Aging: Pathways to Human Life Extension. Springer.
55. Farmanfarmaian, Robin. (2015). The Patient as CEO: How Technology Empowers the Healthcare Consumer. Lioncrest Publishing.
56. Feynman, Richard. (2005). The Pleasure of Finding Things Out: The Best Short Works of Richard P. Feynman. Basic Books.
57. Finch, Caleb E. (1990). Senescence, Longevity, and the Genome. University of Chicago Press.

58. Fogel, Robert William. (2004). The Escape from Hunger and Premature Death, 1700–2100: Europe, America, and the Third World. Cambridge University Press.

59. Fossel, Michael. (2015). *The Telomerase Revolution: The Enzyme That Holds the Key to Human Aging and Will Soon Lead to Longer, Healthier Lives.* BenBella Books.

60. Fossel, Michael. (1996). Reversing Human Aging. William Morrow and Company.

61. Freitas, Robert A. Jr. (2022). Cryostasis Revival: The Recovery of Cryonics Patients thro*ugh Nanomedicine.* Alcor Life Extension Foundation.

62. Friedman, David M. (2007). The Immortalists: Charles Lindbergh, Dr. Alexis Carrel, and Their Daring Quest to Live Forever. Ecco.

63. Fumento, Michael. (2003). BioEvolution: How Biotechnology Is Changing the World. Encounter Books.

64. García Aller, Marta. (2017). El fin del mundo Tal y como lo conocemos: Las grandes innovaciones que van a cambiar tu vida. Planeta.

65. Garreau, Joel. (2005). Radical Evolution: The Promise and Peril of Enhancing Our Minds, Our Bodies, and What It Means to Be Human. Doubleday.

66. Glenn, Jerome, et al. (2018). State of the Future 19.1. The Millennium Project.

67. Gosden, Roger. (1996). Cheating Time. W. H. Freeman & Company.

68. Green, Ronald M. (2007). Babies by Design: The Ethics of Genetic Choice. Yale University Press.

69. Gupta, Sanjay. (2009). Cheating Death: The Doctors and Medical Miracles that Are Saving Lives Against All Odds. Wellness Central.

70. Halal, William E. (2008). Technology's Promise: Expert Knowledge on the Transformation of Business and Society. Palgrave Macmillan.

71. Haldane, John Burdon Sanderson. (1924). Daedalus or Science and the Future. K. Paul, Trench, Trubner & Co.

72. Hall, Stephen S. (2003). Merchants of Immortality: Chasing the Dream of Human Life Extension. Houghton Mifflin Harcourt.

73. Halperin, James L. (1998). The First Immortal. Del Rey, Random House.

74. Harari, Yuval Noah. (2017). Homo Deus: A Brief History of Tomorrow. Harper.

75. Harari, Yuval Noah. (2015). Sapiens: A Brief History of Humankind. Harper.

76. Hawking, Stephen. (2002). The Theory of Everything: The Origin and Fate of the Universe. New Millennium Press.

77. Hayflick, Leonard. (1994). How and Why We Age. Ballantine Books.

78. Hébert, Jean M. (2020). Replacing Aging. Science Unbound.

79. Hessel, Andrew & Webb, Amy. (2022). The Genesis Machine: Our Quest to Rewrite Life in the Age of Synthetic Biology. Public Affairs.

80. Hobbes, Thomas. (2008 [1651]). Leviathan. Oxford World's Classics. Oxford University Press.

81. Hoffman, Rudi. (2018). The Affordable Immortal: Maybe You Can Beat Death and Taxes. Createspace Independent Publishing Platform

82. Hughes, James. (2004). Citizen Cyborg: Why Democratic Societies Must Respond to the Redesigned Human of the Future. Westview Press.
83. Huxley, Julian. (1957). "Transhumanism." New Bottles for New Wine. Chatto & Windus.
84. Immortality Institute (ed.). (2004). The Scientific Conquest of Death: Essays on Infinite Lifespans. Libros En Red.
85. International Monetary Fund. (Annual). World Economic Outlook. International Monetary Fund.
86. Ioviță, Anca. (2015). The Aging Gap Between Species. CreateSpace.
87. Jackson, Moss A. (2016). I Didn't Come to Say Goodbye! Navigating the Psychology of Immortality. D&L Press.
88. Jain, Naveen. (2018). Moonshots: Creating a World of Abundance. Moonshots Press.
89. Kahn, Herman. (1976). The Next 200 Years: A Scenario for America and the World. Quill.
90. Kaku, Michio. (2018). The Future of Humanity: Terraforming Mars, Interstellar Travel, Immortality, and Our Destiny Beyond Earth. Doubleday.
91. Kaku, Michio. (2012). Physics of the Future: How Science Will Shape Human Destiny and Our Daily Lives by the Year 2100. Anchor Books.
92. Kanungo, Madhu Sudan. (1994). Genes and Aging. Cambridge University Press.
93. Kaufmann, Sandra. (2022). The Kaufmann Protocol: Aging Solutions. Independently published.
94. Kennedy, Brian K.; Berger, Shelley, L.; Brunet, Anne; Campisi, Judith; Cuervo, Ana Maria; Epel, Elissa S.; Franceschi, Claudio; Lithgow, Gordon J.; Morimoto, Richard I.; Pessin, Jeffrey E.; Rando, Thomas A.; Arlan Richardson, Arlan; Schadt, Eric E.; Wyss-Coray, Tony & Sierra, Felipe. (2014). "Aging: a common driver of chronic diseases and a target for novel interventions." Cell, 2014 Nov 6; 159(4): pp. 709–713.
95. Kenyon, Cynthia J. (2010). "The genetics of aging." Nature, 464(7288), pp. 504-512.
96. Kuhn, Thomas S. (1962). The Structure of Scientific Revolutions. University of Chicago Press.
97. Kurian, George T. and Molitor, Graham T.T. (1996). Encyclopedia of the Future. Macmillan.
98. Kurzweil, Ray. (2012). How to Create a Mind: The Secret of Human Thought Revealed. Viking Books.
99. Kurzweil, Ray. (2005). The Singularity Is Near: When Humans Transcend Biology. Viking Press Inc.
100. Kurzweil, Ray. (1999). The Age of Spiritual Machines. Penguin Books.
101. Kurzweil, Ray & Grossman, Terry. (2009). TRANSCEND: Nine Steps to Living Well Forever. Rodale Books.
102. Kurzweil, Ray & Grossman, Terry. (2004). Fantastic Voyage: Live Long Enough to Live Forever. Rodale Books.

103. Lents, Nathan. (2018). Human Errors: A Panorama of Our Glitches, from Pointless Bones to Broken Genes. Houghton Mifflin Harcourt

104. Lieberman, Daniel E. (2013). The Story of the Human Body: Evolution, Health, and Disease. Vintage.

105. Lima, Manuel. (2014). The book of Trees: Visualizing Branches of Knowledge. Princeton Architectural Press.

106. Longevity.International. (2017a). *Longevity Industry Analytical Report 1: The Business of Longevity.* Longevity.International.

107. Longevity.International. (2017b). *Longevity Industry Analytical Report 2: The Science of Longevity.* Longevity.International.

108. López-Otín, Carlos; Blasco, Maria A.; Partridge, Linda; Manuel Serrano, Manuel & Kroemer, Guido. (2023). "The Hallmarks of Aging: An Expanding Universe." *Cell,* 2023 January 19; 186(2): pp. 243–278.

109. López-Otín, Carlos; Blasco, Maria A.; Partridge, Linda; Manuel Serrano, Manuel & Kroemer, Guido. (2013). "The Hallmarks of Aging." Cell, 2013 June 6; 153(6): pp. 1194–1217.

110. Maddison, Angus. (2007). Contours of the World Economy 1–2030 AD: Essays in Macro–Economic History. Oxford University Press.

111. Maddison, Angus. (2004). Historical Statistics for the World Economy: 1–2003 AD. OECD Development Center.

112. Maddison, Angus. (2001). The World Economy: A Millennial Perspective. OECD Development Center.

113. Malthus, Thomas Robert. (2008 [1798]). An Essay on the Principle of Population. Oxford World's Classics. Oxford University Press.

114. Martinez, Daniel E. (1998). "Mortality patterns suggest lack of senescence in hydra." Experimental Gerontology, 1998 May;33(3), pp. 217–225.

115. Martínez-Barea, Juan. (2014). *El mundo que viene: Descubre por qué las próximas décadas serán las más apasionantes de la historia de la humanidad.* Gestión 2000.

116. Medawar, Peter. (1952). An Unsolved Problem of Biology. H. K. Lewis.

117. Mellon, Jim & Chalabi, Al. (2017). Juvenescence: Investing in the Age of Longevity. Fruitful Publications.

118. Metzl, Jamie. (2019). *Hacking Darwin: Genetic Engineering and the Future of Humanity.* Sourcebooks.

119. Miller, Philip Lee & Life Extension Foundation. (2005). *The Life Extension Revolution: The New Science of Growing Older Without Aging.* Bantam Books.

120. Minsky, Marvin. (1994). "Will robots inherit the Earth?" *Scientific American,* October 1994.

121. Minsky, Marvin. (1987). *The Society of Mind.* Simon and Schuster.

122. Mitteldorf, Josh & Sagan, Dorion. (2016). *Cracking the Aging Code: The New Science of Growing Old, and What it Means for Staying Young.* Flatiron Books.

123. Moore, Geoffrey. (1995). *Crossing the Chasm: Marketing and Selling High-tech Products to Mainstream Customers.* Harperbusiness.

124. Moravec, Hans. (1999). *Robot: Mere Machine to Transcendent Mind*. Oxford University Press.

125. Moravec, Hans. (1988). *Mind Children*. Harvard University Press.

126. More, Max. (2003). *The Principles of Extropy. Version 3.11*. The Extropy Institute.

127. More, Max & Vita-More, Natasha. (2013). *The Transhumanist Reader: Classical and Contemporary Essays on the Science, Technology, and Philosophy of the Human Future*. Wiley-Blackwell.

128. Mulhall, Douglas. (2002). *Our Molecular Future: How Nanotechnology, Robotics, Genetics, and Artificial Intelligence will Transform our World*. Prometheus Books.

129. Musi, Nicolas & Hornsby, Peter (ed.). (2015). *Handbook of the Biology of Aging, Eight Edition*. Academic.

130. Naam, Ramez. (2005). *More Than Human: Embracing the Promise of Biological Enhancement*. Broadway Books.

131. Navajas, Santiago. (2016). *El hombre tecnológico y el síndrome Blade Runner*. Editorial Berenice.

132. Ocampo, Alejandro; Reddy, Pradeep; Martinez-Redondo, Paloma; Platero-Luengo, Aida; Hatanaka, Fumiyuki; Hishida, Tomoaki; Li, Mo; Lam, David; Kurita, Masakazu; Beyret, Ergin; Araoka, Toshikazu; Vazquez-Ferrer, Eric; Donoso, David; Roman, José Luis; Xu, Jinna; Rodriguez Esteban, Concepcion; Gabriel Nuñez, Gabriel; Nunez Delicado, Estrella; Campistol, Josep M.; Guillen, Isabel; Guillen, Pedro & Izpisua Belmonte, Juan Carlos. (2016). "In Vivo Amelioration of Age-Associated Hallmarks by Partial Reprogramming." *Cell*. 2016 Dec 15; 167(7): pp. 1719–1733.

133. United Nations. (Annual). *Statistical Yearbook*. United Nations.

134. Paul, Gregory S. & Cox, Earl. (1996). *Beyond Humanity: Cyberevolution and Future Minds*. Charles River Media.

135. Perry, Michael. (2001). Forever for All: Moral Philosophy, Cryonics, and the Scientific Prospects for Immortality. Universal Publishers.

136. Pickover, Clifford A. (2007). A Beginner's Guide to Immortality: Extraordinary People, Alien Brains, and Quantum Resurrection. Thunder's Mouth Press.

137. Pinker, Steven. (2018). Enlightenment Now: The Case for Reason, Science, Humanism, and Progress. Viking.

138. Pinker, Steven. (2012). The Better Angels of Our Nature: Why Violence Has Declined. Penguin Books.

139. United Nations Development Programme. (Annual). Human Development Report. United Nations Development Programme.

140. Regis, Edward. (1991). Great Mambo Chicken and the Transhuman Condition: Science Slightly over the Edge. Perseus Publishing.

141. Ridley, Matt. (1995). The Red Queen: Sex and the Evolution of Human Nature. Harper Perennial.

142. Roco, Mihail C. & Bainbridge, William Sims (eds.). (2003). Converging Technologies for Improving Human Performance. Kluwer.

143. Rogers, Everett M. (2003). Diffusion of Innovations. 5th Edition. Free Press.

144. Rose, Michael. (1991). Evolutionary Biology of Aging. Oxford University Press.

145. Rose, Michael; Rauser, Casandra L. & Mueller, Laurence D. (2011). Does Aging Stop? Oxford University Press.

146. Sagan, Carl. (1977). The Dragons of Eden: Speculations on the Evolution of Human Intelligence. Random House.

147. Scott, Andrew; Ellison, Martin & Sinclair, David A. (2021) "The economic value of targeting aging." *Nature Aging* 1, 616–623 (2021)

148. Serrano, Javier. (2015). El hombre biónico y otros ensayos sobre tecnologías, robots, máquinas y hombres. Editorial Guadalmazán.

149. Shakespeare, William. (2017 [1601]). Hamlet. Amazon Classics.

150. Shermer, Michael. (2018). Heavens on Earth: The Scientific Search for the Afterlife, Immortality, and Utopia. Henry Holt and Co.

151. Simon, Julian L. (1998). The Ultimate Resource 2. Princeton University Press.

152. Sinclair, David A. (2019). Lifespan: Why We Age – and Why We Don't Have To. Thorsons.

153. Skloot, Rebecca. (2010). The Immortal Life of Henrietta Lacks. Random House.

154. Stambler, Ilia. (2017). Longevity Promotion: Multidisciplinary Perspectives. CreateSpace Independent Publishing Platform.

155. Stambler, Ilia. (2014). A History of Life-Extensionism in the Twentieth Century. CreateSpace Independent Publishing Platform.

156. Steele, Andrew. (2021). Ageless: The New Science of Getting Older Without Getting Old. Doubleday.

157. Stipp, David. (2010). The Youth Pill: Scientists at the Brink of an Anti-Aging Revolution. Current.

158. Stock, Gregory. (2002). Redesigning Humans: Our Inevitable Genetic Future. Houghton Mifflin Company.

159. Stolyarov II, Gennady. (2013). Death is Wrong. Rational Argumentators Press.

160. Strehler, Bernard. (1999). Time, Cells, and Aging. Demetriades Brothers.

161. Teilhard de Chardin, Pierre. (1964). The Future of Man. Harper & Row.

162. Topol, Eric. (2019). Deep Medicine: How Artificial Intelligence Can Make Healthcare Human Again. Basic Books.

163. United Nations. (2022). World Population Prospects 2022. United Nations.

164. Venter, J. Craig. (2014). Life at the Speed of Light: From the Double Helix to the Dawn of Digital Life. Penguin Books.

165. Venter, J. Craig. (2008). A Life Decoded: My Genome: My Life. Penguin Books.

166. Verburgh, Kris. (2018). The Longevity Code: The New Science of Aging. The Experiment.

167. Vinge, Vernor. (1993). "The Coming Technological Singularity." Whole Earth Review, Winter 1993.

168. Walter, Chip. (2020). Immortality, Inc.: Renegade Science, Silicon Valley Billions, and the Quest to Live Forever. National Geographic.

169. Warwick, Kevin. (2002). I, Cyborg. Century.

170. Weindruch, Richard & Walford, Roy. (1988). The Retardation of Aging and Disease by Dietary Restriction. Charles C. Thomas.
171. Weiner, Jonathan. (2010). Long for This World: The Strange Science of Immortality. HarpersCollins Publishers.
172. Weismann, August. (1892). Essays Upon Heredity and Kindred Biological Problems. Volumes 1 & 2. Claredon Press.
173. Wells, H.G. (1902). "The Discovery of the Future." Nature, 65, pp. 326–331.
174. West, Michael. (2003). *The Immortal Cell.* Doubleday.
175. Wood, David W. (2019). Sustainable Superabundance: A Universal Transhumanist Invitation. Delta Wisdom.
176. Wood, David W. (2018). Transcending Politics: A Technoprogressive Roadmap to a Comprehensively Better Future. Delta Wisdom.
177. Wood, David W. (2016). The Abolition of Aging: The forthcoming radical extension of healthy human longevity. Delta Wisdom.
178. Woods, Tina. (2020). Live Longer with AI. Packt Publishing.
179. World Bank. (Annual). World Development Report. World Bank.
180. World Health Organization. (2011). International Statistical Classification of Diseases and Related Health Problems. 10th Revision, Edition 2010. World Health Organization.
181. World Health Organization. (2006). History of the Development of the ICD. World Health Organization.
182. World Health Organization. (1992). The ICD-10 Classification of Mental and Behavioural Disorders: Clinical Descriptions and Diagnostic Guidelines. World Health Organization.
183. World Health Organization. (1948). Constitution of the World Health Organization. World Health Organization.
184. Young, Sergey (2021). The Science and Technology of Growing Young: An Insider's Guide to the Breakthroughs that Will Dramatically Extend Our Lifespan… and What You Can Do Right Now. BenBella Books.
185. Zendell, David. (1992). The Broken God. Spectra.
186. Zhavoronkov, Alex. (2013). The Ageless Generation: How Advances in Biomedicine Will Transform the Global Economy. Palgrave Macmillan.
187. Zhavoronkov, Alex & Bhullar, Bhupinder. (2015). Classifying aging as a disease in the context of ICD-11. *Frontiers in Genetics.*

Printed in the United States
by Baker & Taylor Publisher Services